The Microbial Cell Cycle

Series editors
Dr J A Cole, University of Birmingham
Dr C J Knowles, University of Kent

Titles in series
Oral Microbiology *P Marsh*
Bacterial Toxins *J Stephen and R A Pietrowski*
The Microbial Cell Cycle *C Edwards*
Bacterial Plasmids *K Hardy*
Extrachromosomal Inheritance *S Oliver*
Respiratory Chain and Photosynthetic Energy Conservation in Bacteria *C W Jones*
Spore Formation in Bacteria *A Moir*
The Structure of the Bacterial Cell *H J Rogers*

Aspects of Microbiology

The Microbial Cell Cycle

Clive Edwards

Lecturer in Microbiology at the University of Liverpool

Nelson

Thomas Nelson and Sons Ltd
Nelson House Mayfield Road
Walton-on-Thames Surrey KT12 5PL

P O Box 18123 Nairobi Kenya

116-D JTC Factory Building
Lorong 3 Geylang Square Singapore 1438

Thomas Nelson Australia Pty Ltd
19–39 Jeffcott Street West Melbourne Victoria 3003

Nelson Canada Ltd
81 Curlew Drive Don Mills Ontario M3A 2RI

Thomas Nelson (Hong Kong) Ltd
Watson Estate Block A 13 Floor
Watson Road Causeway Bay Hong Kong

Thomas Nelson (Nigeria) Ltd
8 Ilupeju Bypass PMB 21303 Ikeja Lagos

© Clive Edwards 1981
First published 1981

ISBN 0 17 771103 5
NCN 5813 42 0

All rights reserved. No part of this publication may be reproduced,
stored in a retrieval system, or transmitted, in any form or by any means,
electronic, mechanical, photocopying, recording or otherwise,
without the prior permission of the publishers.

Phototypeset by Tradespools Ltd, Frome, Somerset
Printed and bound in Hong Kong

Contents

1 Introduction	**1**
2 Methods for investigating the cell cycle	**5**
Preparation of synchronous cultures	5
Induction methods	8
Selection methods	15
Culture fractionation	22
Naturally occurring synchrony	23
References	24
3 The DNA-division cycle	**25**
Prokaryotic DNA-division cycle	26
Eukaryotic DNA-division cycle	40
References	47
4 RNA and protein synthesis	**48**
RNA synthesis	49
Protein synthesis	52
Models of microbial enzyme synthesis	53
Other control mechanisms	60
References	63
5 Growth, differentiation, and respiration during the microbial cell cycle	**64**
Prokaryotic differentiation	64
Respiration in prokaryotes	67
Surface extension and the prokaryotic cell cycle	69
Eukaryotic cell cycle events: the use of yeasts as experimental material	73
Biogenesis of mitochondria in eukaryotes	77
References	81
6 Overall summary	**82**
Further reading	84
Glossary	**84**
Index	**87**

Preface

The purpose of this book is to show that the behaviour of populations of micro-organisms in batch cultures rarely reflects that of individual cells. I did not intend it to be a comprehensive summary of the state of current knowledge of the microbial cell cycle. Instead, I have attempted to provide the background to the main areas of the subject and, more importantly, to compare the cycles of prokaryotes and eukaryotes. There are omissions, but I have sought to cover the gaps by giving pertinent references, especially to review articles. The final chapter contains a personal selection of currently popular areas of study.

I am indebted to Dr J.A. Cole and Dr C.J. Knowles for their helpful comments during the early stages of this work, to Mrs J. Johnston and Mrs L. Marsh for preparation of the typescript, and to my wife Tisha for cheerfully helping with the correction of proofs.

1981   CLIVE EDWARDS

1 Introduction

The main aims of this book are to show that cellular events occur in sequences which can be revealed only by studying cells at known stages in the cell cycle; to review the methods which can be utilized to unravel cell cycle processes; to give examples of some of the major cellular processes which have been investigated—DNA and protein synthesis, for example; to demonstrate by means of commonly used systems how micro-organisms are being utilized to study various aspects of their cycle; and to attempt to point out some of the main differences and similarities which are apparent in the cell cycles of prokaryotic and eukaryotic micro-organisms.

A useful starting point is to look at the various terminologies commonly encountered in the literature. The *cell cycle* can be defined as a fixed period during which a cell, newly formed at division, grows until it divides to form two daughter cells. The duration of the cycle will depend upon the particular cell type and the prevailing environmental conditions. For example, a bacterium such as *Escherichia coli* can have a cycle time of 20 min in a complex rich medium whereas in simpler salts media containing a known energy source the cycle will be longer, assuming all other conditions such as temperature remain the same.

During each cycle, new cell material must be synthesized so that, at division, each daughter cell receives an equal portion of all components. It will be obvious that the cellular complement must at least double during the cycle otherwise it becomes diluted in successive generations. This synthesis of new materials must be integrated and regulated in an *ordered sequence* at the correct locations within the cell. Also, each cell component must be present at the appropriate time in the cell cycle. The cycle therefore has *spatial* and *temporal* controls which constitute a differentiation process at the cellular and molecular levels.

Traditionally, the various cellular processes of micro-organisms have been investigated in *batch culture*—that is, growth is exponential for a few generations until nutrients become depleted and then gradually ceases. *Continuous cultivation* of micro-organisms has also been employed, using *chemostats* (in which growth depends on the rate of supply of a limiting nutrient) or *turbidostats* (in which culture density governs the flow-rate of incoming nutrients). Both approaches attempt to standardize growth conditions and maintain the organisms at a fixed growth rate. The populations of both batch and continuous cultures will consist of cells at all stages of the cell cycle ranging from young newly divided cells to older cells at the point of division. Such populations, therefore, can be said to consist of *asynchronously dividing cells*. Any event measured in these cells, such as bacterial wall synthesis, will reflect only a *time-averaged* value over the whole of the cell cycle. Newly divided bacteria may synthesize cell wall more rapidly than those at division but measurements of this parameter in asynchronous cells gives only a mean value and tells us nothing about patterns of synthesis during the cycle. To obtain further information about the sequence of events, the investigator must be able to obtain cells at a known stage of the cycle. Two approaches that can be used to achieve this involve studying single organisms or populations which are at the same stage of the cell cycle.

Measurements of cell cycle parameters in single cells are severely limited by the range of techniques that can be used. Biochemical determinations usually require much more experimental material. Parameters which can be assessed by microscopic examination include increases in volume and cell growth. This technique can be very accurate for those cells which grow only in length. A good example is the fission yeast *Schizosaccharomyces pombe*. It has been shown by microscopical measurements that an increase in cell volume of this yeast occurs only over the first three-quarters of the cell cycle.

Because of the obvious limitations of single-cell methods and the ever-increasing development of new biochemical techniques, a large number of methods have been devised for obtaining populations of cells which are of similar age and hence at the same stage in the cell cycle. One approach is to prepare *synchronous cultures*, in which cells are of similar age and which, therefore, grow and divide synchronously. Another approach is to separate cells from an asynchronous culture into age classes which, collectively, represent the whole cell cycle. This latter technique is called *culture fractionation*. These two approaches will be examined in greater detail in the next chapter. Suffice to say that none of the methods employed to obtain cells of similar age is free from criticism. More importantly, they are not universally applicable, since some of the methods have been developed for specific cell types: for example, light/dark cycles for inducing synchrony in photosynthetic organisms.

A whole range of cell cycle events can be measured once a system has been developed. Unfortunately, there appears to be a gulf between many areas of research and the use of cell cycle methods such as synchronous cultures. Perhaps the chief cellular component that has been studied during the cell cycle of both prokaryotes and eukaryotes is DNA synthesis. Other important processes, such as membrane biogenesis, cellular respiration and associated enzymes, have been sadly neglected until recently. This is surprising in view of the known temporal stages undergone during DNA synthesis in both prokaryotes and eukaryotes. These stages are more apparent in eukaryotes since DNA synthesis is always periodic, occurring at a particular stage of the cell cycle known as the *S-phase* (synthetic phase). The integration of this event with other cellular processes has not been extensively studied.

Synthesis of various enzymes during the cell cycle of both prokaryotes and eukaryotes can occur continuously or periodically during the cycle. Patterns of synthesis vary for different enzymes or even for the same enzyme in different cell types. Measurements of enzyme levels during the cell cycle are fraught with difficulties. An assumption widely made in the literature is that enzyme activity is directly proportional to enzyme amount; unfortunately this may be an oversimplification. Another factor which influences interpretation of enzyme activity during the cell cycle is that of *enzyme potential*. Observed enzyme activity need not reflect the maximum possible activity at a given stage in the cycle. This problem is best illustrated by examples of inducible enzymes such as ß-galactosidase. In the repressed state very little enzyme activity will be present but this may increase a thousandfold in the fully induced state. Other parameters affecting enzyme activity *in vitro* include the presence of intracellular proteases and loss or dilution of an essential co-factor during assay. Unfortunately these considerations are not always taken into account in studies of patterns of enzyme synthesis during the cell cycle.

Both prokaryotic and eukaryotic micro-organisms can usually be cultured conveniently and rapidly in large amounts in easily prepared growth media. They provide a whole spectrum of cellular processes, each of which can differ between species. This diversity and ease of cultivation makes micro-organisms particularly suitable for investigating cell cycle events.

The choice of a particular organism will depend on a variety of criteria—the ease of establishing it in synchronous culture for example—but it is usually influenced by a particularly interesting property of the physiology of the organism. Rod-shaped bacteria such as *Bacillus subtilis* have been used extensively to study surface growth characteristics during the cell cycle. Other morphologically differentiating bacteria such as *Caulobacter crescentus* are suitable for investigating flagellar biogenesis and synthesis of phage attachment sites during the cycle. In suitable culture media, pigmented prokaryotes such as *Rhodopseudomonas palustris* can be utilized to further our understanding of bacterial photophosphorylation when they are grown in the light, and of oxidative phosphorylation when they are grown in the dark. This particular prokaryote has the added bonus of a dimorphic cell cycle giving rise to a swarmer cell and a sedentary tubed cell. Other bacteria are useful in cell cycle studies because of their extreme adaptability and ease of genetic manipulation. A noteworthy example of such a genetically well-defined prokaryote is *Escherichia coli*.

Eukaryotic micro-organisms also provide a variety of cell types which have been utilized in studies of the cell cycle. Yeasts have been used extensively in this respect, especially *S. pombe*, which is readily established in synchronous culture by a variety of methods. Another commonly used yeast is *Saccharomyces cerevisiae*. The yeasts make useful systems for investigating respiratory development during the cycle because they can be grown fermentatively or non-fermentatively. Protozoa such as *Tetrahymena pyriformis* and the amoebae, which tend to be susceptible to lysis after relatively mild disruption procedures, make suitable cell systems for following the development of intra-cellular organelles during the cycle. Unicellular algae such as *Chlorella* can be utilized to investigate photosynthetic reactions.

These are potentially exciting experimental systems for increasing our knowledge of cellular physiology. However, enthusiasm should be tempered by an awareness of the danger of applying an understanding of a process in one organism to that in another. Micro-organisms are notoriously wayward entities which can carry out what is superficially the same process *via* different pathways or by utilizing different enzymes. This complexity may be manifest in a single organism and not confined to inter-species differences. A good illustration of this is electron transport in aerobic bacteria: a single bacterium such as *E. coli* can oxidize reduced substrates using different pathways of electron flow and respiratory carriers, depending on the prevailing growth conditions.

2 Methods for investigating the cell cycle

Every stage of the cell cycle will be represented in a growing asynchronous culture. Yet some events (such as DNA synthesis) are confined to discrete periods of the cycle. Hence, in order to study such events it becomes necessary to work with cells of similar age and follow them through the phases of their cycles. Unfortunately, this is not achieved simply. We shall see that the methods used to prepare cultures of micro-organisms homogeneous with respect to age are rarely free from criticism. The main objection is that they perturb the treated cells, giving rise to artefacts and abnormal cell cycle events. Also, certain species of micro-organisms are suitable only for specific methods for obtaining cells of different ages. Perhaps the best example is the regime of alternating light and dark periods which induces synchronous division of photosynthetic micro-organisms. In this chapter we will concentrate on methods of preparing synchronous cultures, the separation of asynchronous cells into age classes by culture fractionation, and naturally occuring synchrony.

Preparation of synchronous cultures

Over the past few years many techniques have been introduced for preparing cultures of synchronously dividing micro-organisms. Some work well only for one type of organism whereas others have wider applications. The purpose of these methods is to yield a population of cells which is relatively homogeneous with respect to age and which grows and divides synchronously.

Assessment of the extent of synchrony A synchronous culture is one in which all the cells are at the same stage of the cell cycle, so that they grow and divide together. Such a culture amplifies the phases undergone by a single cell. Theoretically, all the cells of a synchronous culture should grow and divide at the same time, giving rise to a stepwise increase in cell numbers. In practice, this is never achieved: the division period is prolonged because of slight inherent age differences between cells (Fig. 2.1). The shorter the division period, the higher the degree of synchrony. Various indices have been developed to assess how close particular synchronous cultures come to the ideal or theoretical case. Such indices also allow us to compare the degree of synchrony in different experiments and to determine the efficiency of the various methods used to achieve synchronous

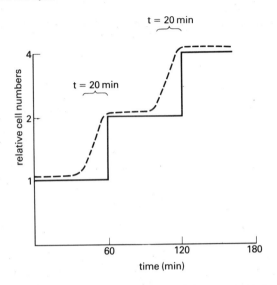

Fig. 2.1 Comparison of idealized synchronous growth (———) with that obtained in practice (– – – – –).

growth. It is important to use the same index for all experiments because application of different indices can give quite different results.

An often used index is that of Blumenthal and Zahler (1962), which is given by

$$F = N/N_0 - 2^t/g$$

where F is the synchrony index, N_0 and N are the numbers of cells before and after division respectively, t is the time in minutes taken for division, and g is the *generation time* (the time in minutes for all the components of the culture to double in amount). This index takes into account both the extent to which doubling has been achieved (ideally N/N_0 should equal 2.0) and the time taken for a synchronous burst of division compared with the generation time of cells during exponential growth.

For a perfect synchronous culture F will be 1.0. For the experimental culture in Fig. 2.1, the generation time is 60 min, the time taken for division is 20 min, and $N/N_0 = 2.0$. This gives an F value of 0.75. An asynchronous culture with a doubling time of 60 min would have an F value of 0.0 as assessed by the above index. In practice, an index between 0.5 and 0.7 indicates an acceptably high degree of synchrony.

Balanced growth A central dogma of investigations using synchronous cultures is that the organisms should exhibit *'balanced growth'*. This requires that all measurable parameters, such as cell numbers, dry weight and respiratory activity, should increase by the same amount over the same time interval. Thus, in every cell cycle, a doubling of all the components should be observed. Because of discontinuities of synthesis of some cell constituents during the cell cycle (DNA replication, for example) the time taken for one cycle to be completed is the minimum interval that can be used to assess the extent of balanced growth.

Techniques developed for preparing synchronous cultures can be grouped into either *selection methods* or *induction methods*. These have been reviewed by Mitchison (1971) and Helmstetter (1969), respectively. Normally, as we shall see later, selection methods give rise to fewer metabolic disturbances and hence more 'balanced growth' than do induction methods. However, much information may be gleaned by deliberately perturbing the normal cell cycle using an induction method or by drastically modifying the growth conditions. Furthermore, adequate control experiments should always be carried out under conditions as near as possible to those used to prepare a synchronous culture. These should include measurements of the desired parameter in a culture in the exponential phase of growth as well as in a culture treated in a similar fashion to that used to obtain a synchronous culture. For example, cells separated according to size on sucrose gradients should be re-mixed to form the original asynchronous population. Unfortunately some methods, such as those which call for the use of inhibitors, are not amenable to adequate control experiments.

Presentation of data Because of the slight variation in division times of synchronously growing cells, it is important to be able to delineate the cell cycle in a reproducible fashion. Usually, the duration of the cycle is estimated from the mid-point of the first division to the mid-point of the second, and represented as a linear scale 0–1.0. Measurements of timing of events during the cell cycle are related to these points and adjusted to fall on the linear scale. For example, in Fig. 2.2a, oxygen uptake rates are shown to oscillate, rising to two maxima per cell cycle. From the mid-point of the first division to the mid-point of the second gives a cycle time of 75 min. We assign the first mid-point as 0 on our linear scale and the second as 1.0. Thus, each 0.1 unit of the linear scale is equal to 7.5 min. The first peak in the oxygen uptake rate occurs 25 min after the first mid-point and, therefore, at 0.33 of a cycle. The second peak in the oxygen uptake rate occurs 60 min after the first mid-point and, therefore, at 0.80 of a cycle. These results can be summarized as a cell cycle map, which is

The microbial cell cycle

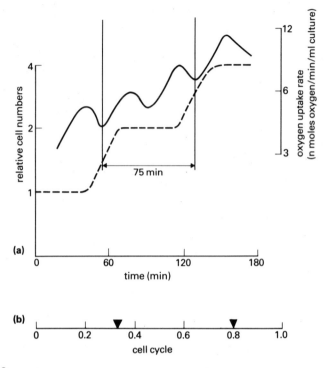

Fig. 2.2
a Theoretical pattern of oxygen uptake rates (———) during synchronous growth of a micro-organism (– – – –).
b Cell cycle map summarizing the timings of the maxima of oxygen uptake rates during the cycle. The cell cycle is represented as a linear scale from 0 to 1.0 which corresponds to the time from the first to the second mid-point of increase in cell numbers.

also shown in Fig. 2.2. By repeating the experiment a number of times, all timings of maxima of oxygen uptake rates can be represented on the map and the mean calculated. This method of presentation compensates for small variations in cycle times between successive experiments, and cycle maps provide a standard method of representing periodic events.

Induction methods

Induction methods involve the treatment of asynchronously dividing cells over a period of time in order to bring them into the same phase of the cell cycle. Some induction methods rely on deliberately

Methods for investigating the cell cycle

perturbing or shocking the culture and throw serious doubt on the normality of the resulting synchronous division and cell cycle events.

Induction based on end-points of growth Micro-organisms continue to grow only as long as their nutritional requirements are met. Once an essential growth factor is depleted, growth stops. This can be observed during the transition from exponential growth to the stationary phase of a batch culture. Often more than one growth factor will have been depleted and it is difficult to identify which is the first factor to become limiting. However, for some bacteria, resuspension of cells from the end of exponential growth into fresh growth medium causes a synchronous pattern of division for two or three cell cycles.

This method does not have universal application: it is successful for only a few species such as *B. subtilis*, *E. coli*, *Proteus vulgaris*, and some yeasts. The precise point at which organisms can be removed and resuspended in fresh growth medium to initiate synchronous growth is important. As can be seen from Fig. 2.3, cells in the late

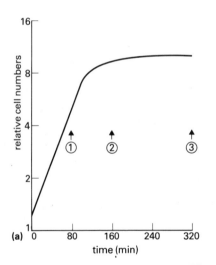
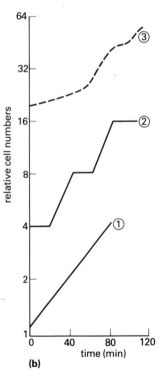

Fig. 2.3 The effect of transferring cells at different stages of asynchronous growth to fresh growth medium.
a Growth curve of a culture from which 1, exponential cells, 2, early stationary phase cells, and 3, late stationary phase cells are transferred to fresh growth medium.
b Subsequent patterns of growth of samples 1, 2, and 3.

exponential phase of growth usually continue to divide exponentially when resuspended in fresh nutrients, whereas cells at the onset of stationary phase (culture 2 in Fig. 2.3) divide synchronously. Major criticisms of this method are that the resuspended cells exhibit an abnormally long cell cycle which is poorly defined—that is to say, the exact mechanism by which cells are arrested by nutrient depletion is not known.

Another starvation method employed for obtaining synchronous division of *Bacillus* species is based on the synchronous germination of spores. This method probably works because the spores represent a homogeneous population which are similar in composition and hence will germinate synchronously. Little work has been done to demonstrate that events occurring during synchronous spore germination reflect those which occur during the cycle of vegetative cells.

The resuspension of stationary-phase cells into fresh growth medium and the synchronous germination of spores are examples of poorly defined methods for studying the cell cycle because neither the cause of synchrony nor the effect on the subsequent cell cycle is known. Other starvation methods which have more defined effects rely on induction of cell synchrony by the removal of *one* known essential growth factor from an exponential culture. However, the method succeeds only if the growth of each individual cell stops at the same stage of the cell cycle. Perhaps the commonest example of this approach is amino acid deprivation of amino acid auxotrophs. The essential amino acid is withdrawn for approximately one generation time and then added back to the culture. After re-supplementation, synchronous growth will occur only if *each cell was arrested at the same point in the cell cycle during the withdrawal period.*

Vitamin deprivation, such as the deprivation of thiamine from cultures of the flagellate protozoon *Polytomella agilis*, has been used to prepare synchronous cultures. An unusual specific starvation regime is used to prepare synchronous cultures of diatoms. These organisms require silicon to form the shell which encloses the cells. Silicon deprivation for a fixed period from a culture of the diatom *Cylindrotheca fusiformis*, followed by its restoration, has been employed successfully to prepare synchronous cultures of this organism.

Induction by hypoxia The deprivation of oxygen (*hypoxia*) from an asynchronous culture of an aerobic organism, followed by its reintroduction, can induce some micro-organisms to divide synchronously. This approach has been used to induce synchrony in cultures of the protozoon *Tetrahymena pyriformis*. The technique involved the application of hypoxia for a period of 4 h, causing cells to accumulate in the G2 phase of the cell cycle (see page 40, Chapter 3). When aeration was resumed, synchronous division occurred. However, it must be appreciated that such a method can induce profound physiological changes, especially with strictly aerobic micro-organisms. For example, it is known that under such conditions micro-organisms undergo changes in their respiratory chains and hence their capacity for energy production is also modified.

Induction by inhibitors Various chemicals can block asynchronously growing cultures at definite stages of the cell cycle. If they are exposed to the inhibitor for at least a generation time, then theoretically *all* the cells of that culture will progressively become blocked at the same stage of the cell cycle. If the inhibitor is then replaced by fresh growth medium, synchronous growth will commence. A good inhibitor for induction of synchronous cultures must be specific for a definite phase of the cell cycle; must block *all cells immediately* they reach this stage of the cycle; must be removed readily by washing procedures; and must not damage or kill a proportion of the culture.

Commonly used inhibitors are those which block DNA synthesis. These are especially suitable for eukaryotic micro-organisms in which DNA synthesis is confined to the S-phase (see page 40, Chapter 3). One such inhibitor, hydroxyurea, has found wide application in inducing synchronous growth in mammalian cells, and in micro-organisms such as the trypanosomatid *Crithidia luciliae*, and in the yeast *S. pombe*. Unfortunately, this inhibitor may be selectively lethal during the S-phase of some organisms so that, on its removal, the culture contains a proportion of non-dividing cells.

Deoxyadenosine is of interest because it inhibits DNA synthesis without greatly affecting RNA synthesis or increases in dry weight. It has found wide application in mammalian cells and micro-organisms. In the fission yeast *S. pombe*, it has been used by treating a growing asynchronous culture with 2mM deoxyadenosine for 3 h, slightly longer than the usual 2.5 h doubling time of this organism (Mitchison and Creanor, 1971). On removal of the inhibitor, there is a burst of DNA synthesis followed by a synchronous division. Dry weight and RNA synthesis continue during treatment causing greater variation in size and mean size of cells before the first division than is the case in asynchronous cultures.

These observations have led to the concept that there are two cycles—the *DNA division cycle* and the *growth cycle*. Normally both are linked, but when DNA synthesis becomes blocked with deoxyadenosine, the DNA-division cycle is inhibited while the growth cycle (RNA, dry weight) continues. This dislocation of the two cycles leads to an abnormal size distribution before the synchronous division that occurs after removal of deoxyadenosine. On addition of the inhibitor, those cells in S-phase will be inhibited immediately, whereas those which have just completed S-phase will continue to grow and divide until they again reach S-phase. The latter class of cells will be approximately the usual size for this stage, but the former will continue to grow during the arrest period and become abnormally large. Therefore, treatment with deoxyadenosine will synchronize the DNA-division cycle but *not* the growth cycle. This ability to dissociate the two cycles provides a powerful manipulative tool for comparing events during 'normal' cell cycles in selection-synchronized cells in which both the DNA-division cycle and growth cycle are synchronous with those in the deoxyadenosine-induced synchronous DNA-division cycle.

Induction by periodic supply of growth factors This approach is a refinement of the technique for inducing synchrony by starvation. Methods have been developed, largely with chemostat cultures, for periodically pulsing an essential growth factor to an asynchronous culture. The important requirements are that the time between each pulse ideally should be the normal generation time for the given growth medium, and that the growth factor pulsed is usually enough to maintain growth for one doubling. Most methods rely on periodic pulsing of the carbon source as the limiting growth nutrient but other growth factors such as various amino acids have also been used.

The method has been used to induce synchronous growth in a strain of *E. coli* (Buckley and Anagnostopoulos, 1975). Briefly, the method involves growing the bacteria in salts medium containing a limiting amount (0.25 g per litre) of glucose. Starting with an asynchronous culture of 100 ml, 50 ml of fresh growth medium are added every cycle time, which may be more, less, or the same as the generation time of cells during growth in batch culture. On addition of fresh medium, 50 ml of the total volume (150 ml) are removed, thus maintaining a constant volume of 100 ml. Continuous cycling in this manner eventually leads to synchronous division of the cells, ascribed to a blockage of DNA replication during the pulsing treatment. Further work is required to check the validity of this method, since it is possible that only the DNA-division cycle is synchronous, while the growth cycle is not. It is not known whether the events occurring during periodic pulsing are the same as those occurring in cell cycles of synchronous cultures established by other methods, especially selection methods. The technique has the attractive feature of altering the duration of the cell cycle by varying the time between pulses, and also of establishing a continuous synchronous culture system.

Methods for investigating the cell cycle

Induction by temperature changes Periodic changes in temperature provided one of the first techniques for inducing synchronous division in micro-organisms. Cultures at optimal growth temperature are heated or cooled rapidly to temperatures unfavourable to growth and held there for a fixed period of time. The method is most successful if the shifts are repeated at intervals equal to the generation time in batch growth. This approach is illustrated by a regime developed to induce synchrony in the protozoon *Tetrahymena pyriformis*. A culture of the protozoon is grown at 28°C (optimal growth temperature); every generation time (155–160 min) the temperature is raised to 34°C (division-blocking temperature) for 0.5 h. After five or six heat shocks the culture divides synchronously.

The method has serious drawbacks. It can be used only for small culture volumes which equilibrate rapidly when the temperature is altered. Subjecting the cells to a growth-blocking temperature for 0.5 h may lead to serious metabolic disturbances—indeed, *Tetrahymena* cells may become two to four times their normal size after the heat shock. Similar temperature changes have been used to induce synchronous division in *Salmonella typhimurium*, in protozoa such as *T. pyriformis*, and in algae such as *Astasia longa*. This method is of limited application today because of the criticisms outlined above.

Induction by alternating light and dark cycles These methods are applicable only to photosynthetic micro-organisms. The principle is

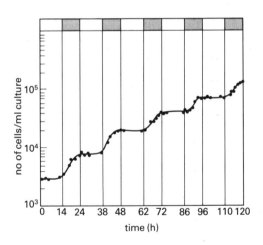

Fig. 2.4 Light-induced synchrony in *Euglena gracilis*. Cell numbers increase synchronously through five divisions. Shaded squares (10 h duration) represent dark periods; unshaded (14 h duration) light periods. (Redrawn from Lor & Cossins, 1973.)

illustrated by a light/dark regime used to synchronize the photosynthetic unicellular organism *Euglena gracilis* (Lor and Cossins, 1973). At 25°C, with sufficient illumination, *E. gracilis* will grow with an average generation time of 24 h. If alternate light and dark periods of 14 h and 10 h respectively are introduced, a small percentage of cells will be able to divide in the dark period. As the light/dark cycles are repeated, however, this percentage will increase until all the cells of the culture come into phase and grow and divide together (Fig. 2.4). The disadvantage of this technique is that it is suitable only for low culture densities.

Light cycles are commonly employed to induce synchronous cultures of a variety of unicellular algae such as *Chlorella vulgaris* and *Chlamydomonas reinhardii*. Some have been modified to operate as a continuous synchronous culture. Interestingly, the non-photosynthetic protozoon *T. pyriformis* has been induced to divide synchronously by light/dark treatment of a late exponential phase culture. This may possibly be caused by the production of a light-sensitive pigment during the late exponential phase of growth.

Summary of induction methods

Induction methods generally rely on treating an asynchronously dividing population of cells over a period of time, as a result of which that population grows and divides together. Such methods usually gain high yields of synchronously dividing cells. However, most of the methods described cause appreciable metabolic disturbances—indeed, they rely on these perturbations to induce synchronous growth.

Generally, induction methods affect specific stages of the cell cycle and are of two types. *Single-shock methods* rely on treatment of an asynchronous culture with one shock which is cumulative—that is to say, all the cells are gradually arrested at a defined point in the cell cycle. On removal of the inducing stimulus, synchronous growth occurs. Examples of such methods include hypoxia, the use of inhibitors, and transfer of stationary phase cells into fresh growth medium.

In contrast, *periodic shocks* are applied at fixed intervals, usually one generation time apart. They have the effect of gradually bringing an asynchronous population of cells into the same phase. In this category are included periodic provision of growth nutrients, temperature changes, and the light/dark transitions employed to obtain synchronous cultures of photosynthetic micro-organisms.

Selection methods

Whereas induction methods involve treatment of a population of asynchronous cells, selection methods rely on the physical separation of an age or size class of cells from an asynchronous culture. Generally, these techniques rely on the size alteration which normally occurs during the cell cycle: a cell emerging from cell division will be smaller and lighter than a cell about to undergo division. The main source of perturbation will be the means by which a particular age class is selected. Although it is generally accepted that synchronous cultures prepared by selection methods are more suitable for studying the stages undergone by a single cell during its cell cycle, such cultures can at best be termed only *minimally perturbed* since they will have been subjected to some separation treatment. Selection methods can be broadly divided into those based on *filtration* and those based on *centrifugation*.

Selection by filtration There are many types of filtration method which have largely been developed in conjunction with improved filters and filtration apparatus. Probably one of the first to be used was developed for *E. coli*. It involves the concentration of an exponential culture of the organism by centrifugation to form a thick suspension, which is then drawn by suction into the upper layers of a stack of filter papers held in a Buchner-type funnel. Growth medium is then poured on to the pile and drawn through by suction, drawing the cells from the upper layers through the stack of filter papers. The upper layers retain the largest cells while the lower layers contain the smallest cells of the original culture. By resuspension of a filter paper from either the upper or lower layers into fresh culture medium, synchronous growth can be initiated. The yield of either large or small cells is approximately five per cent of the original culture. The method has obvious drawbacks. Concentration of the cells by centrifugation removes them from growth medium, causes anaerobiosis, and subjects the organisms to rapid sedimentation. These factors may cause metabolic shocks which in turn would cause abnormal cycles. The procedure is also fairly time-consuming up to the point of resuspension.

Successive workers have introduced modifications to overcome these criticisms. The initial centrifugation step has been omitted so that the asynchronous culture is passed directly into a stack of filters, the cells being retained. Further passage of fresh pre-warmed growth medium gradually elutes (washes out) the smallest cells, which can then be used as the source of a synchronous culture. Even with this modification the cells are still removed from growth medium and resuspended in fresh growth medium. This potentially perturbing

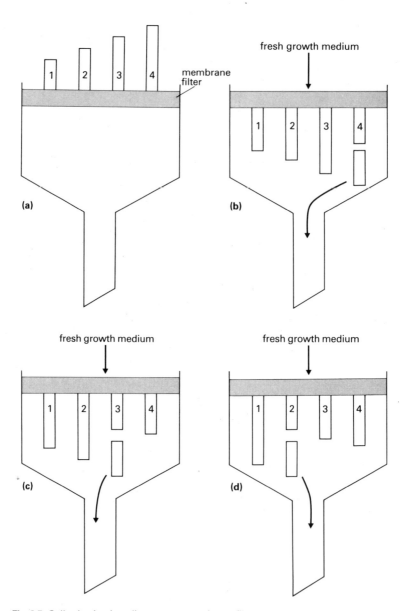

Fig. 2.5 Cell selection by adherence to membrane filters.
a Asynchronous cells are drawn on to the filter (shown stippled) by suction. Cell sizes range from newly-divided cells (1) to cells about to divide (4).
b The filter is inverted and pre-warmed growth medium is drawn gently through it. The cells continue to grow and divide (cell 4) on the underside of the filter.
c Cells of size similar to cell 3 divide and, in **d**, size 2 cells divide. Periodic collection of fractions ensures a relatively homogeneous suspension of cells of similar age.

Table 1.1 Evaluation of filtration methods

Advantages	Disadvantages
Relatively mild procedures which can be performed at growth temperatures.	May introduce perturbations if the initial asynchronous population is concentrated before filtration.
Usually rapid and technically undemanding.	A degree of trial and error is involved in setting up a filtration method for a particular micro-organism; even an established system can give varying results.
No need for expensive apparatus.	
	Most filtration methods give very low yields.

treatment can be overcome by judicious choice of filters so that between five and ten per cent of the population passes directly through the filters when an asynchronous culture is poured on to them. These cells should be the smallest ones and will have remained in their original growth medium during the treatment. If the selection is done at the growth temperature in an incubator room, it is relatively gentle and provides a minimally perturbing regime.

The final modification of selection by filtration which we will consider here exploits the adherence of some cells to membrane filters. Asynchronous cells are concentrated on to a membrane filter of pore size sufficient to restrict the passage of all the cells through it. The filter is then turned over and growth medium is gently and continuously pumped through it so that the cells continue to grow on the underside of the filter (Fig. 2.5). Newly divided cells will not be attached to the filter and will be eluted in the growth medium. Removal of the eluate after a predetermined period yields a suspension of similar-aged cells which can be used to initiate synchronous growth.

Selection by centrifugation: rate-separation in density gradients This method was first developed for *E. coli* by Mitchison and Vincent (1965). An asynchronously dividing population of cells is concentrated rapidly by centrifugation and resuspended in a small amount of growth medium. The concentrated suspension is then layered on to a sucrose gradient contained in a tube. The sucrose is dissolved in growth medium to minimize perturbations to the cells, and the gradient is prepared so that the density increases linearly with the depth of the tube. All procedures, including centrifugation, should be carried out at a temperature as close to the growth temperature of the organism as possible. After layering the cells on the less dense top of

The microbial cell cycle

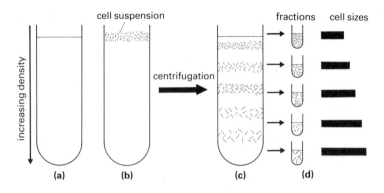

Fig. 2.6 Rate separation of a cell suspension in a density gradient.
a Tube containing a gradient of a medium such as sucrose whose density increases with depth.
b A concentrated cell suspension is carefully layered on to the less dense surface of the gradient.
c The tube is centrifuged so that the cells become distributed through the gradient according to size. Smaller cells remain in areas of low density near the top of the tube; heavier cells travel into the gradient.
d The gradient is progressively withdrawn in fractions of equal volume which contain cells of similar size. Fractions from the lighter end of the gradient will contain small cells; those from areas of high density, larger cells. Fractions can be inoculated into fresh growth medium to initiate synchronous growth.

the gradient, the tube is centrifuged long enough for the cells to become distributed along most of the length of the gradient. Large cells will travel furthest and be located near the bottom of the tube in regions of high density, but the smallest will remain near the top (Fig. 2.6).

The gradient can now be divided into a number of fractions, each of which will contain organisms predominantly at the same stage in the cell cycle. Thus, synchronous growth can be initiated by inoculating growth medium (pre-warmed to the growth temperature) with a suitable number of fractions containing similar-sized cells. In practice, the inoculum should not contain more than ten per cent of the original cell population and, usually, fractions containing the smallest cells are used.

The method has certain disadvantages:

> Cells are concentrated by centrifugation before separation.
>
> Anaerobiosis may be accentuated during the separation procedure and perturbations may arise because of the osmotic effects of sucrose, especially at high density. This can be offset to some extent by always using small cells from low density sucrose.

Methods for investigating the cell cycle

Temperature changes occur.

The method takes an appreciable amount of time, probably 15–30 min from concentrating cells to collecting fractions. This has obvious disadvantages when working with micro-organisms which have cell cycle times of 40–90 min.

Despite these disadvantages this method has been successfully employed to prepare synchronous cultures of a variety of micro-organisms ranging from *E. coli* to yeasts such as *S. pombe* and the protozoon *Crithidia fasciculata*. The osmotic effect of sucrose can be overcome by using an isotonic gradient material such as ficoll or dextran. Another drawback that occurred when this method was first devised was the low yields of similarly aged cells obtained from density gradients in tubes. However, the development of zonal rotors has overcome this.

The method employing zonal rotors is basically the same as that for separations in tubes except that the density gradient is set up in a hollow rotor. The concentrated cell suspension is pumped into the centre of the spinning rotor, which has been pre-filled with the density gradient (low density near the centre, gradually increasing to high density at the rotor wall). The rotor is accelerated from the loading speed to the separating speed for a time sufficient to sediment cells out from the centre into the gradient. Fractions are removed by pumping, and those containing the smallest cells are used to inoculate fresh medium for synchronous growth. Thus, more cells can be separated using this scaled-up procedure.

All of the criticisms raised against separations in tubes apply equally to those in rotors. Indeed, it is probable that the first synchronous division which occurs after inoculation of the fractions containing the smallest cells into growth medium is abnormal. It is imperative, therefore, when using synchronous cultures prepared by this method to follow at least two synchronous divisions.

Selection by centrifugation: continuous-flow centrifugation The inherent disadvantages outlined for selection by density gradients have been largely overcome by a recently developed method based on size-selection by continuous-flow centrifugation (Lloyd *et al.*, 1975). Continuous-flow centrifuges were originally developed to concentrate micro-organisms from large volumes of culture medium. This is achieved by passing liquid culture continuously through a rotor spinning at a sufficiently high speed to sediment the cells from the growth medium on to the rotor wall, with the cleared medium passing out of the rotor to waste. A diagram showing how this centrifugal method can be adapted to select for a size class of cells is shown in Fig. 2.7.

The microbial cell cycle

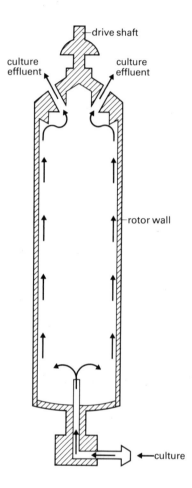

Fig. 2.7 Principles of size-selection using continuous-flow centrifugation. Culture is passed into the base of the spinning hollow rotor at a predetermined flow-rate and rotor speed. Arrows show the movement of the culture vertically up the rotor wall and out through to waste *via* exit ports at the top of the rotor. Most cells are sedimented onto the rotor wall, large ones at the bottom and progressively smaller ones near the top. Those escaping sedimentation pass out in the culture effluent and can be used to initiate synchronous growth.

The asynchronous culture, aerated and maintained at the growth temperature, is siphoned into the spinning continuous-flow rotor. The rate of flow of culture into the rotor can be adjusted, as can the rotor speed, so that nearly all the cells are sedimented. However, between five and fifteen per cent are allowed to escape sedimentation and pass out of the rotor in the culture effluent. These represent the lightest and smallest cells of the original culture. Recent work with *E. coli* has suggested that the selected cells may, in fact, not be the smallest cells but, as long as a small enough percentage of cells is allowed to escape sedimentation, the resulting suspension seems to represent a definite stage of the cell cycle. Therefore, the rotor

effluent contains an age class that can be incubated and assessed for synchronous growth. The continuous-flow method overcomes nearly all the disadvantages of other selection methods.
The advantages are:

Cells are maintained at a constant temperature and are never removed from the growth medium.

The procedure is rapid. Cells are subjected to centrifugal forces for seconds rather than for several minutes as in gradient methods. Because high yields of synchronously dividing cells can be obtained rapidly, the method is particularly suitable for micro-organisms with short cycle times. A flow-rate of 300 ml min^{-1} has been used to obtain synchronous cultures of *Alcaligenes eutrophus*.

It is a relatively mild procedure and has been used successfully to prepare synchronous cultures of fragile organisms such as the amoeba *Acanthamoeba castellanii*.

Oxygen deprivation does not occur because the asynchronous culture can be aerated before it is passed through the rotor, and the selected cells in the rotor effluent can be aerated immediately on collection. The time spent within the rotor is negligible.

The method is suitable for any non-filamentous, non-aggregating organism.

Size selection using continuous-flow centrifugation has been employed successfully to prepare synchronous cultures of a variety of micro-organisms. These range from bacteria such as *E. coli* and *A. eutrophus* to yeasts such as *S. pombe* and *Candida utilis* and protozoa such as *T. pyriformis* and *A. castellanii*.

Selection by centrifugation: counter-flow centrifugation This method is sometimes referred to as *centrifugal elutriation*. In this procedure the centrifugal force—that is, the force which sediments particles in a centrifugal field—is counter-balanced by a liquid flowing in the opposite direction. A hollow centrifuge rotor rather similar to a zonal rotor is used, and the counter-flow liquid is forced inwards from the rotor rim towards the centre, while the particles to be separated are propelled out from the centre to the rotor rim. By judicious choice of flow-rate, the heavier faster-sedimenting cells are 'allowed' to overcome the force of the counter-flowing liquid and are collected as a fraction. By successively lowering the flow-rate of the counter-flow liquid, a series of fractions of progressively lighter and lighter cells can be collected. Some of these fractions are used to inoculate growth medium and the ensuing culture is monitored for synchronous growth. This method does not yet have wide application but it has been used to prepare synchronous cultures of the yeast *Saccharomyces cerevisiae*.

Summary of selection methods

The preparation of synchronous cultures by selection methods usually involves the separation of a size or age class from an asynchronously growing culture. This may be achieved:

By using filtration methods which perturb the treated culture minimally; however, filtration methods generally suffer from lack of reproducibility, and give low yields of synchronously dividing cells.

By using centrifugal methods to select small cells from an asynchronous culture by separation on density gradients or by size-selection using continuous flow or counter-flow centrifugation.

Selection methods are usually preferred to induction methods because it is generally accepted that they produce synchronously dividing cells which more closely reflect the division cycle of a single cell. However, even selection methods may perturb the treated culture and, therefore, any synchronous culture prepared by such methods can at best be minimally perturbing.

Culture fractionation

Another way in which the cell cycle may be represented as a series of fractions, each of which contains a similarly aged class of cells, is by use of a technique known as *culture fractionation*. The methods used stem largely from the size separations achieved using either density gradients or continuous-flow centrifugation.

Density gradient method An asynchronous population is rapidly concentrated by centrifugation and layered on to a density gradient contained either in a centrifuge tube or a zonal rotor. After centrifugation for a few minutes, the cells will be distributed throughout the gradient according to size. Small cells about to begin the cycle will be near the least dense top part of the gradient; the largest cells, which will have reached the end of their cycles and are ready to divide, will be located in denser parts of the gradient. By dividing the gradient into fractions, an ordered series ranging from small cells through to large cells and representing all the stages of the cell cycle can be obtained. As long as the micro-organism used has an easily measurable parameter which is closely linked to a known stage of the cell cycle, each fraction can be assigned to a definite stage.

The advantage of this approach is that by scaling-up the procedure in zonal rotors, large amounts of cells at known stages of the cell cycle can be selected, allowing for large-scale biochemical determinations. Culture fractionation on density gradients has been applied successfully to study enzyme activities during the cycle of the fission yeast *S. pombe* (Poole and Lloyd, 1973). Fortunately in this yeast, cell volume increases almost linearly with distance travelled by the cells from the rotor centre over the first three-quarters of the cell cycle. This fraction of the cell cycle, therefore, can be readily resolved in this organism after the culture fractionation procedure.

Continuous-flow centrifugation The principles of this method have already been outlined. Instead of allowing a percentage of cells to escape sedimentation, all the cells of a culture are pelleted on to the rotor wall. The largest cells will be located near the bottom of the rotor (Fig. 2.7), followed by progressively lighter cells up the rotor wall. Usually a rolled up plastic sheet is inserted to line the rotor wall. After centrifugation the sheet can be removed and the pelleted cells collected layer by layer as a series of fractions. As in the density gradient method, the organism used must have an easily identified parameter which occurs at a known stage of the cell cycle in order to assign the fractions to particular stages of the cell cycle.

Summary

Culture fractionation is a rapid method for splitting a population of asynchronous cells into age classes. Furthermore, large numbers of cells can be isolated at definite stages of the cell cycle. It is important that any event occurring in a particular age class isolated during culture fractionation should also be apparent at the same stage of the cell cycle during synchronous growth. When used together, synchronous cultures and culture fractionation methods provide powerful tools for examining cell cycle events, particularly those that occur periodically during the cycle.

Naturally occurring synchrony

In a number of biological systems cellular events occur synchronously without any treatment by external agents. Perhaps the most well known are the initial synchronous divisions undergone by a fertilized egg, the synchrony of which decays after two or three divisions. Outgrowth of bacterial spores of some *Bacillus* species may also be synchronous, but the biochemical events occurring during this process cannot be compared readily with those in a growing vegetative cell.

The most studied naturally occurring synchrony in microorganisms is that found in the acellular slime mould *Physarum polycephalum*. This eukaryote has a complicated life-cycle consisting of unicellular stages and a plasmodial stage. The plasmodium may contain thousands of nuclei but is completely acellular. The nuclei undergo synchronous divisions as the plasmodium grows. Methods have been developed by which single large plasmodia 30 cm in diameter can be grown aseptically. Although nuclear division is synchronous there is, of course, no accompanying cell division.

References

BLUMENTHAL, L.K. & ZAHLER, S.A. (1962). 'Index for measurement of synchronization of cell populations.' *Science* 135: 724.
BUCKLEY, D.E. & ANAGNOSTOPOULOS, G.D. (1975). 'Growth of *Escherichia coli* B/r/1 in a semi-continuous system designed for the synchronization of cell division.' *Archives for Microbiology* 105: 169–172.
HELMSTETTER, C.E. (1969). 'Methods for studying the microbial division cycle.' *Methods of Microbiology* 1: 327–361.
LLOYD, D., JOHN, L., EDWARDS, C. & CHAGLA, A.H. (1975). 'Synchronous cultures of micro-organisms: large-scale preparation by continuous-flow size selection.' *Journal of General Microbiology* 88: 153–158.
LOR, K.L. & COSSINS, E.A. (1973). 'One-carbon metabolism in synchronized cultures of *Euglena gracilis*.' *Phytochemistry* 12: 9–14.
MITCHISON, J.M. (1971). *Biology of the Cell Cycle*. Cambridge University Press, London.
MITCHISON, J.M. & CREANOR, J. (1971). 'Induction synchrony in the fission yeast *Schizosaccharomyces pombe*.' *Experimental Cell Research* 67: 368–374.
MITCHISON, J.M. & VINCENT, W.S. (1965). 'Preparation of synchronous cell cultures by sedimentation.' *Nature* 205: 987–9.
POOLE, R.K. & LLOYD, D. (1973). 'Oscillations of enzyme activities during the cell cycle of a glucose-repressed fission yeast. *Schizosaccharomyces pombe* 972 h⁻. *Biochemical Journal* 136: 195–207.

3 The DNA-division cycle

During the cell cycle a cell must double its mass before division, replicate its genome, and ensure an even partitioning of genetic material to the two daughter cells. In this chapter we will look at the way prokaryotes and eukaryotes synthesize DNA and how this vital process is linked to cell division and growth. First we will concentrate on prokaryotes, especially *E. coli*. An outline of similar events in eukaryotic micro-organisms is then presented, mainly for comparison.

One of the most perplexing and fascinating areas of biology is differentiation (how cells become specialized to carry out specific functions). Examples of cell differentiation range from the morphological transitions between stalked and swarmer cells of the prosthecate bacteria of the genus *Caulobacter* to the transition of amoeboid cells of the slime mould *Dictyostelium discoidium* to stalk and spore cells and to the complex and spectacular differentiation of fertilized eggs into complex life forms.

Much time and effort is being devoted to discovering how and why certain cells become triggered to differentiate at a particular time. The controls involved must be both *temporal* (occurring at different times) and *spatial* (occurring at different sites). Many investigators have argued that because differentiation by such complex systems as developing embryos cannot be understood at the present time, simpler systems should be studied in the hope that this will lead to new insights which can be extrapolated to the more complicated systems. This reductionist approach is valid only if one remembers that not all cell types carry out similar functions in the same way. As we shall see, although there are many similarities between DNA synthesis in prokaryotes and in eukaryotes, there are also major differences. At best then, information gained from one system can serve only as a model for comparison with other systems.

The process of replication of DNA has been studied extensively in both prokaryotic and eukaryotic micro-organisms. Most of the work on prokaryotes has been confined to *E. coli*, and reservations about generalizing the *E. coli* models to other prokaryotes must remain foremost in our minds. Happily, studies of DNA synthesis of eukaryotic micro-organisms have been more diverse, enabling direct comparisons to be made.

Prokaryotic DNA-division cycle

The emphasis in this area on *E. coli* is due largely to our broader understanding of the molecular and genetic processes of this bacterium. This rather one-sided view of the prokaryotic world may well lead to misconceptions until other bacteria become better characterized. Fortunately, there is also an ever-growing understanding of the DNA-division cycle of *Bacillus subtilis*.

Chromosome replication The most important event of the cell cycle is that of chromosome replication. Ever since the elucidation of the structure of DNA by Watson and Crick, much time and effort has been spent in investigating how such a large and complex molecule is replicated. Furthermore, we must ask how the synthesis of this macromolecule is synchronized with cell growth and division. Two basic approaches have been employed. One involves the identification and characterization of the individual elements responsible for replication. The other involves understanding the control systems of effectors and repressors of DNA synthesis, and how these processes fit in with other cell cycle events. These approaches may be summarized as the study of *DNA synthesis* and the *regulation of DNA synthesis*. We will not deal with the elements responsible for DNA synthesis here; suffice to say that they include DNA polymerases, DNA ligating enzymes (proteins which join ends of DNA), single-stranded DNA-dependent ATPase, RNA polymerase, and DNA unwinding enzymes. We will concentrate largely on the relationship of DNA synthesis to cell division, control of replication and, in particular, the temporal aspects of the process.

The physical characteristics of the *E. coli* chromosome have recently been reviewed by Kleppe *et al.*, (1979). The molecule consists of double-stranded DNA organized into a single circular chromosome approximately 1100 μm long when unfolded. There is also evidence that the chromosome of *B. subtilis* is circular. The *E. coli* chromosome lies free in the cell apart from at least one attachment point to the cytoplasmic membrane.

It has been proposed that the cell cycle can be divided into three biochemically distinct periods during slow growth (James, 1978). The first period has been called the *I-period* and its duration depends on how rapidly the cells are growing. During rapid growth the I-period will be of short duration or even absent, whereas it will be extended during slow growth. This period is one of preparation for the beginning of chromosome replication. Various processes must be achieved so that the cell reaches a critical time in the cycle when it is ready to *initiate chromosome replication*.

The second period, called the *C-period*, is the phase of the cycle

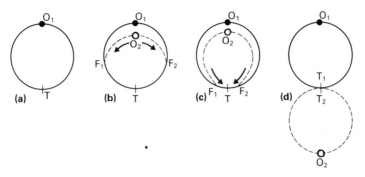

Fig. 3.1 Bidirectional chromosome replication.
a Chromosome replication is initiated at the origin O_1 of the circular chromosome.
b and **c** Once initiated, replication proceeds *via* two replication forks (F_1 and F_2) which move in opposite directions until they reach the terminus T. As a result, **d**, two sister chromosomes (———— and - - - - -) are produced, each having an origin (● and ○) and a terminus (T_1 and T_2).

during which chromosome replication takes place. In *E. coli* growing with doubling times between 20 and 70 min, the C-period is a fixed 40 min duration. Chromosome replication is semi-conservative and begins at a definite site called the *origin* (Fig. 3.1). Once initiated, replication proceeds bidirectionally away from the origin, the actual point of replication being termed a *replication fork*. Thus, two replication forks are initiated at the origin and move in opposite directions round the circular chromosome, eventually meeting at a common point called the *terminus*. In *E. coli*, the rate of replication at each fork is approximately 14 μm of chromosome replicated per min. The rate of replication is, therefore, independent of the growth rate, at least in cells growing with doubling times between 20 and 70 min.

The third period of the cycle has been termed the *D-period* and lasts for 20 min in *E. coli*. The D-period is the time between the termination of chromosome replication (that is, when both replication forks have reached the terminus) and cell division.

This means that the sum of the C and D-periods in *E. coli* is a fixed duration of 60 min. Evidence for this has come from measurements of the rate of DNA synthesis after asynchronous cultures have been pulse-labelled with tritiated thymidine. Label is incorporated only into those cells that are actively synthesizing DNA. The cells are removed from the labelled medium and drawn onto a membrane filter. The filter is then everted and growth medium is pumped through at a low flow-rate so that the adhered cells continue to grow on the underside of the filter. (This experimental system has also been used to prepare synchronous cultures, as shown in Fig. 2.5.) The oldest cells divide and their progeny are eluted into the growth medium, whereas those cells furthest from division continue to grow on the filter until eventually, after almost one cell cycle time, they too divide and their progeny are eluted. By collecting samples of the eluate over a doubling time for a particular set of conditions, fractions of cells representative of the different stages of the cell cycle are obtained.

These fractions represent the progeny of cells which divided soon after adherence to the membrane through intermediate stages to cells derived from those which were furthest from division at the time of adherence. If DNA synthesis is continuous throughout the cell cycle, all the progeny in the fractions will be labelled equally. In fact, a twofold jump in the specific activity of labelled thymidine (expressed as radioactive counts/cell) is observed at a definite stage of the cell cycle. This implies that the rate of DNA synthesis doubles at a definite point in the cell cycle which corresponds to a doubling of

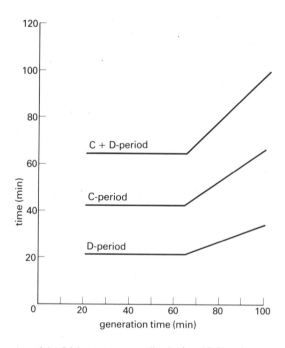

Fig. 3.2 Duration of the C (chromosome replication) and D (time between the termination of chromosome replication and cell division) periods in *E. coli* growing at different doubling times. (Redrawn from Helmstetter, 1969.)

The DNA-division cycle

initiation sites and, hence, to the beginning of the C-period. By repeating the experiment with cells growing at different doubling times (and therefore different cell cycle times) the durations of the C and D-periods have been calculated and plotted against cell cycle times. As can be seen from Fig. 3.2, C and D times remain constant in those cells growing with doubling times between 20 and 70 min.

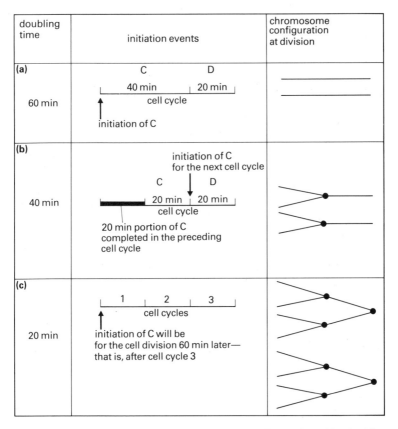

Fig. 3.3 Initiation of chromosome replication in *E. coli* cells growing with a doubling time of 60, 40, or 20 min.
a Initiation of C occurs at or just after cell division. The single parallel lines represent the replicated circular chromosomes at the end of the D period.
b Initiation of C occurs 20 min into the previous cell cycle—that is, at the beginning of D. The closed circles represent replication forks which at division are already half way round each chromosome.
c Initiation of C is for a cell division 60 min later (C = 40 min + D = 20 min), which in this case occurs three generations later.

Within these limits, therefore, *the rate of chromosome replication in E. coli is independent of growth rate.*
This presents us with a problem. How does *E. coli* growing rapidly in broth cultures with doubling times of 20–30 min replicate its chromosome, a process which apparently takes 40 minutes? This paradox has been resolved by the discovery of *multiple replication forks.* By such means, initiation of a new round of chromosome replication can occur before the previous one has been completed. It seems that prokaryotes can be completing several cycles of chromosome replication at once (Fig. 3.3). A cell growing with a doubling time greater than 60 min (that is, $60 + x$ min) will have an I-period of x minutes before initiation of chromosome replication. Cells growing with doubling times of 60 min or less have no I-period.

A cell growing with a doubling time of 60 min, for example, initiates chromosome replication during or immediately after division. Chromosome replication lasts for 40 min and is followed by a 20-min D-period and then by cell division. In a cell growing with a doubling time of 40 min, initiation of chromosome replication will have occurred 20 min into the previous cell cycle. At division, therefore, the replication forks will already be halfway around the chromosome. The round will be completed in 20 min and cell division takes place after the D-period of 20 min. At the onset of the D-period, a new round of replication is initiated for the following cycle. The effect of cells dividing more and more rapidly, therefore, is to push initiation of chromosome replication further into preceding cycles. In cells with doubling times of 20 min, the initiation of replication corresponds to a cell division three generations later.

An important point arising from chromosome replication in rapidly growing cells is that of *gene dosage.* The genes nearest the origin will be present in greater copy number than those genes near the terminus because of multi-forked replication. *Control of chromosome replication is independent of growth rate, therefore, and depends on the frequency of initiation.* It is apparent that, if the assumption that the C and D-periods are constant is correct, then an initiation event in *E. coli* will be followed by cell division 60 min later. Cell division is therefore probably controlled at initiation of chromosome replication.

Regulation of chromosome initiation The discovery that the frequency of chromosome initiation governed the rate of chromosome replication rather than changes in the rate at which the replication forks traversed the chromosome led to the proposal that initiation occurred when the cell reached a certain size—in other words at a *critical cell initiation mass.* This is often denoted as *Mi.* Indeed, it has been shown that cell mass at initiation is constant for cultures of *E.*

The DNA-division cycle

coli growing with doubling times between 30 and 60 min, and twice as much for cells with doubling times between 20 and 30 min.

Many models have been proposed to explain this relationship. Among the first was the *replicon model* which arose out of studies of the F plasmid. This plasmid is present in low copy number per cell but exhibits extremely stable inheritence. In order to account for this Jacob *et al.* (1963) proposed the replicon model, which postulates that there are specific attachment sites for the plasmid on the bacterial membrane. Growth of membrane between these sites ensures an ordered segregation of sister plasmids at cell division.

This model has been adapted to explain chromosome replication and segregation in *E. coli*. It is proposed that initiation of DNA synthesis is a membrane-dependent event and is controlled by gene products. A structural gene produces an initiator protein (or proteins) which acts on the membrane-associated origin to initiate chromosome replication. By the time chromosome replication is terminated, the membrane will have grown between the two sister origins, each of which will remain bound to its membrane site. Therefore, not only does the membrane participate in initiating

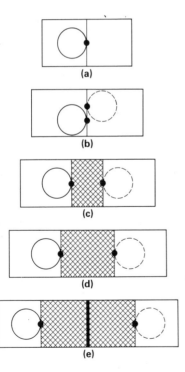

Fig. 3.4 The replicon model for attachment and segregation of DNA by the membrane in *E. coli*.
a A single chromosome is attached at one point to the membrane.
b Replication of the genome brings about the formation of a sister chromosome (dotted line) and duplication of membrane attachment sites.
c and **d** Subsequent insertion of membrane (cross-hatched area) between the sites segregates the chromosomes to opposite poles of the cell, ensuring, **e**, that each cell arising from division possesses a copy of the genome. Note that these events need not occur in this strict sequence. New membrane can be inserted immediately after duplication of attachment sites and during the chromosome replication period. (Redrawn from Jacob *et al.*, 1963.)

replication, it also serves to segregate the chromosomes towards opposite poles of the cell in readiness for cell division (Fig. 3.4). The replicon model proposes a function for the bacterial membrane analagous to that of the mitotic apparatus of eukaryotes. Evidence in support of the replicon model has come from observations that the chromosomes of *E. coli* and *B. subtilis* are indeed membrane-associated.

Effector and repressor proteins Recent work on the inheritance of the F plasmid has shed some doubt on the validity of the replicon model. Perhaps one major criticism is that it is difficult to envisage a permanent chromosome-binding site on the membrane. Neither does the replicon model satisfactorily account for the constancy of mass at initiation. It is difficult to imagine that a cell has some kind of spatial awareness—that it can titrate its mass so that it knows when to initiate replication. More recently, it has been proposed that there is either an effector or an initiator protein whose concentration parallels cell mass; as cell mass increases, so does the concentration of the initiator until it reaches a critical concentration (at Mi) which is sufficient to trigger chromosome replication; alternatively there may be a repressor (possibly a protein) of chromosome replication whose concentration is progressively diluted as cell mass increases until it decreases sufficiently to allow chromosome replication to be initiated.

Evidence for the importance of protein synthesis in DNA initiation has come from studies in which protein synthesis was inhibited by chloramphenicol, or by starvation of an auxotroph of *E. coli* for an essential amino acid. These studies indicated that although incomplete rounds of replication were completed after protein synthesis had been inhibited, *no new rounds were initiated*. This suggests that protein synthesis is not required for DNA elongation; the proteins involved in the process have already been synthesized and are stable enzymes. In contrast, any effector protein must be unstable because no new initiations occur; conversely, any putative repressor must be stable.

Pritchard (1978) has presented an appealing argument for the existence of a repressor of initiation of chromosome replication. He proposed that a repressor-like protein binds at a region near the origin of replication. The synthesis of repressor is linked to initiation so that each initiation event leads to a burst of production of repressor molecules. Thus, the initiation event is closely coupled to an immediate increase in the concentration of the repressor molecule. This reduces the probability of new initiation occurring at the site of the newly replicated origin of the future sister chromosome and fits in with a critical cell mass at initiation. The theory predicts that repressor molecules are synthesized periodically, and that the timing of synthesis is closely coupled to initiation. Once synthesized, the repressor molecule becomes progressively diluted as the cell grows. Under steady state conditions, this simple regulation mechanism would maintain a constant ratio of chromosome origins per unit volume of cytoplasm.

This repressor model for the regulation of initiation of chromosome replication also predicts that the burst of repressor molecule production must occur almost instantaneously, in concert with the replication of the origin, and its concentration must rise to at least twice that found before initiation, in order effectively to prevent initiation of the resultant sister origins after chromosome replication has been triggered. A hypothetical representation of repressor concentration at successive initiation events is shown in Fig. 3.5. An

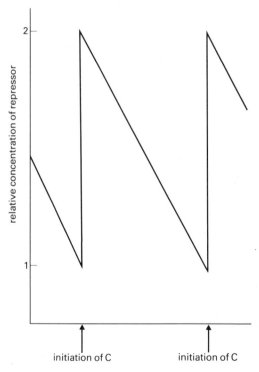

Fig. 3.5 Postulated rise in concentration of a repressor molecule at initiation of chromosome replication (C) followed by its subsequent dilution to a concentration which allows the next initiation event. (Redrawn from Pritchard, 1978).

alternative proposal to a burst of synthesis of stable repressor molecules is the continuous synthesis of an unstable repressor during the cell cycle.
To date, no effector or repressor proteins have been identified. This could be because they are extremely short-lived molecules or because they are present at very low concentrations in the cell. This latter consideration is important in view of the very small area of the chromosome at which initiation must occur. Cellular concentrations of effector or repressor proteins may be so low that they are undetectable by present methods. Current investigations are approaching this problem using temperature sensitive (*ts*) mutants in which DNA is synthesized at a growth temperature (such as 30°C) but is immediately blocked in the chain elongation stage at a nonpermissive temperature (such as 42°C) or in which chain elongation is unaffected by shifting to the non-permissive temperature but no new rounds of chromosome replication are initiated.

Role of ATP in chromosome replication of *E. coli* Respiratory inhibitors such as carbon monoxide and potassium cyanide inhibit chromosome replication in *E. coli*. This suggests that replication is highly energy-dependent (Forterre *et al.*, 1972). It is now thought that there are many ATP-dependent reactions involved in chromosome replication. Various DNA-dependent ATPases have been identified, (DNA gyrase, for example), which seem to require single-stranded DNA as a co-factor for hydrolysing ATP. These enzymes can attach to DNA in the absence of ATP but, in doing so, undergo a configurational change which reveals the ATP-hydrolysing site. It is proposed that on ATP hydrolysis, a further configurational change occurs which leads to movement of the enzyme and possibly of the DNA itself.

Recent work has revealed that the *E. coli* chromosome is not randomly coiled within the cytoplasm. Certain areas or domains are supercoiled. Also, a 'histone-like' protein has been isolated which is closely associated with specific regions of the chromosome. This leads to the intriguing possibility that not all of the chromosome is available for transcription. Progression of a replicating fork round the chromosome might occur in the following stages:

1 elimination of chromosomal superstructures (for example, supercoiling) to form a template for the polymerization complex. This stage includes unwinding of the DNA duplex *via* a DNA-dependent ATPase;
2 polymerization of the nucleotide triphosphates to the newly synthesized strands;
3 reformation of chromosomal superstructures (now two chromo-

somes) behind the replicating fork *via* ATP-dependent DNA gyrase.

Thus, hydrolysis of ATP is an essential component of the replication process. It appears that if steps 1 and 3 are inhibited by nalidixic acid or by the absence of ATP *in vitro*, a limited portion of the chromosome is still replicated. However, inhibitors of step 2 (for example, azide or mitomycin C) completely block DNA synthesis. A possible control of DNA replication by the intracellular concentrations of the adenine nucleotides is an interesting possibility, especially since periodicities of respiratory events have been observed in *E. coli* and *A. eutrophus*. Any depletion of intracellular ATP would seriously affect chromosome replication. Methods for establishing synchronous cultures of micro-organisms by starvation may work because of a gradual depletion of ATP.

Summary

Our understanding of chromosome replication in prokaryotes stems largely from work with *E. coli* and, in part, *B. subtilis*. The cell cycle can be divided into I, C and D-periods, which are biochemically different. The duration of the I-period depends on how rapidly cells divide, whereas that of the C and D-periods are fixed at 40 and 20 min, respectively. The chromosome is replicated during the C-period and is initiated at the origin *via* two replication forks which move bidirectionally round the circular chromosome to meet at the terminus. The rate of DNA replication depends upon the frequency of initiation and not on changes in the rate at which the chromosome is replicated. Cell division occurs 20 min after the termination of chromosome replication. Initiation is an important point of regulation. It is not known how initiation of replication is regulated but it may involve control *via* effector and/or repressor molecules which reach a critical concentration at a critical cell mass, which is termed the initiation mass. Protein synthesis is required for the initiation event but as yet no initiator protein has been isolated. The process of chromosome replication is highly energy-dependent.

Chromosome segregation Once a cell has replicated its genome it is faced with the problem of segregating the sister chromosomes so that, at division, each cell contains a complete copy. Prokaryotes regularly and accurately achieve this process without the apparent involvement of a complicated structure such as the mitotic apparatus of eukaryotes. The bacterial cell envelope has been implicated in this process. In *E. coli* the cell envelope is made up of a cytoplasmic or inner membrane which contains many enzymes (such as those of the respiratory chain, permeases, and enzymes for synthesizing the outer layers), and a rigid peptidoglycan layer which is important in maintaining cell shape and withstanding the osmotic pressures inside the cell as well as the processes of growth and division. This rigid layer must also allow for the location of other surface structures such as antigens. There is also an outer membrane which is more rigid than the inner membrane and contains fewer protein species. The major protein in this layer is lipoprotein, some of which is attached to the murein. Normally, an increase in cell volume during the cell cycle is due to the growth and elongation of the surface layers. This usually occurs in one dimension. For example, rod-shaped bacteria such as *E. coli* and *B. subtilis* increase in length while the cell diameter remains constant. However, at division, the cell surface must be extended in a second dimension in order to form a cross-wall.

A role for the cell wall in bacterial growth and division has been demonstrated in many ways, including the use of mutants defective in one or more functions, which result in morphological or other alterations in these processes. Perhaps the most striking example of the function of the cell wall in growth and division comes from investigations with *B. subtilis*. When this bacterium is treated with lysozyme in an osmotically protective buffer, the cells lose their walls and form spherical bodies called protoplasts. On transfer of protoplasts to an appropriate solid medium they form colonies called *L-forms* which contain cellular units of widely differing sizes and shapes. Morphological and other studies of L-forms suggest that the precise and ordered events which result in segregation of sister chromosomes no longer occur. The process is disorganized and haphazard, and some cellular units lack DNA or other essential cellular components. When the L-forms are returned to conditions suitable for wall synthesis, regular and ordered division will occur only after a morphologically recognizable wall has formed and a rod shape has been re-created. These results imply a role for the peptidoglycan layer of *B. subtilis* in initiating chromosome replication, cell enlargement, and chromosome segregation.

Three models which have been proposed for chromosome segregation are represented in Fig. 3.6. The first is based on the replicon model of Jacob *et al.* (1963) and proposes that replication is initiated at a specific point—the origin—which is in close association with a definite membrane site. The membrane is proposed to provide the enzymes and the regulatory elements for initiation, replication, and termination of DNA synthesis, and to segregate the chromosomes by growth between the attached sister chromosomes. This model requires the permanent attachment of chromosomes to the membrane at all stages of the cell division cycle. The second model proposes that new surface growth occurs to one side of the

The DNA-division cycle

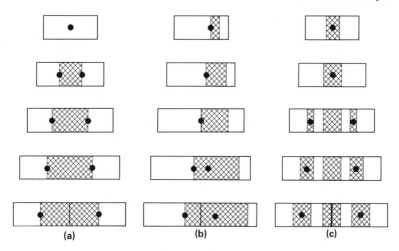

Fig. 3.6 Hypothetical models of chromosome segregation in bacteria.
a The replicon model, in which insertion of membrane between replicated chromosomal attachment sites segregates the chromosomes (see Fig. 3.4).
b Membrane biogenesis occurs to one side of the chromosome only, which has a permanent attachment site on the membrane.
c Once replicated the chromosomes migrate to opposite poles. This model requires the generation of new membrane attachment sites. After attachment, membrane is synthesized on both sides of the chromosome. (Redrawn from Sargent, 1974.)

chromosome and again requires permanent attachment of chromosomes to the membrane. The third model does not require permanent attachment but proposes that the chromosome is attached to a surface structure during replication. On completion of replication, the sister chromosomes dissociate from their attachment sites and migrate to two new surface sites at which replication can be initiated. As in the previous two models, the membrane is required to generate new sites.

There is good evidence for chromosome attachment to the cytoplasmic membrane in *B. subtilis*. Also, the series of enzymes required for chromosome replication have been shown to be closely associated with membrane. Gentle lysis of *E. coli*, followed by sucrose density centrifugation, yields DNA associated with fragments of the cell envelope. Treatment of these complexes with phospholipase leads to release of DNA from the complex, implicating phospholipids in the attachment of the chromosome. An intriguing question regarding chromosome segregation is whether segregation is a random or non-random process. Does one chromosome tend to segregate towards the same pole of the cell during successive

generations? This is extremely difficult to answer experimentally but some notable attempts have been made.
In *E. coli*, chromosome segregation is a non-random process, indicating a relationship between the pole towards which a DNA strand segregates in one generation and the pole to which it segregates in the following division. Two models have been proposed: 1, permanent association of DNA with a pole of the cell such that the sister strand will segregate non-randomly; and 2, distinction between the strands—any association of a strand at a pole of the cell is *not permanent*. Variability in the degree of non-randomness (as may occur in different growth media, for example) is a result of a change in the strength of association between a strand of DNA and a particular pole of the cell. Model 1 requires a permanent chromosome attachment site, for which there is little experimental evidence. The best that can be said at present is either that chromosome segregation is essentially a non-random process but superimposed on it are events which can result in a degree of randomness or, alternatively, chromosome segregation is a random process which has an element of non-randomness superimposed on it (Cooper *et al.*, 1978).

Chromosome replication and cell division Any model of cell division must take into account the pattern of chromosome replication during the cell cycle, the dimensions of the cell, and any changes in these parameters which may occur at different growth rates. A model for the events of the C and D-periods leading to cell division is presented in Fig. 3.7. It is proposed that alongside the initiation of chromosome replication a separate protein synthesizing sequence is activated. Proteins synthesized during the C-period will be involved in cell division. At termination of chromosome replication it is postulated that the synthesis of *termination proteins* is triggered which, once synthesized, will help trigger the division proteins at the end of the D-period. It appears that the D-period can be sub-divided into two distinct phases which have been denoted D1 and D2. During D1 there is extensive macromolecular synthesis, while during D2 the bacteria are committed to divide and no DNA, RNA, or protein synthesis is required for division.

The model proposes, therefore, that the cell division cycle of prokaryotes consists of two coupled cycles, one of which consists of chromosome replication and cell division and the other, which has been termed the *growth cycle*, consists of macromolecular synthesis. Evidence for the two cycles comes from the observation that inhibition of chromosome replication in rod-shaped bacteria may result in continuation of the unaffected growth cycle so that the cells continue to grow as filaments.

The DNA-division cycle

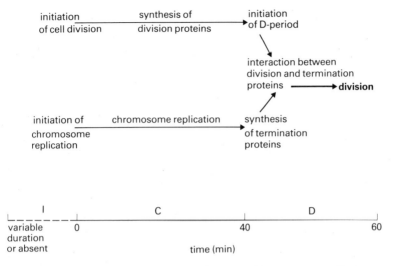

Fig. 3.7 Postulated events occurring during the *E. coli* DNA-division cycle. (Redrawn from Whittenbury, 1978).

Regulation of cell septation In *E. coli* it is known that termination of DNA replication is an essential pre-requisite to cell septation. Nalidixic acid is an inhibitor of DNA elongation—that is, it inhibits chromosome replication. When this inhibitor is added at different times during synchronous growth it is effective only in blocking cell division until 20 min before it occurs. If added later (during the D-period) there is no effect on cell division. During the D-period the cell envelope of *E. coli*, instead of growing lengthways, must be triggered to invaginate to form a septum. This requires the co-ordinated re-direction of all three layers. The inner cytoplasmic membrane and murein layers of the envelope invaginate first, followed by the outer membrane. The signal for re-directing the surface layers inwards at a *specific time in the cell cycle* is not understood. It is not known whether the septum murein differs from that of the cylindrical cell walls.

Attempts to understand this process better have involved the use of ß-lactam antibiotics. Low concentrations of ß-lactams such as penicillin G inhibit septation in *E. coli*, resulting in long multinucleate filaments. High concentrations cause lysis. These two effects may reflect septum–specific murein synthesis which can be prevented by low concentrations of penicillin, or murein synthesis for cell elongation which can be inhibited only by high concentrations of penicillin. Another approach to investigating the mechanism of regulation of septum–specific murein synthesis is by measurement of the activity of enzymes known to have a role in murein biosynthesis. Any enzyme which exhibits a periodic variation in activity corresponding to the time of septation could be implicated in the process. However, as we shall see in the next chapter, one cannot necessarily correlate enzyme *activity* with enzyme *amount*.

Summary

Chromosome segregation is thought to require the participation of the cell envelope, in particular the cell membrane. Specific membrane attachment sites are proposed which may be permanent or be newly synthesized every cell cycle. Experimental evidence suggests that segregation of sister chromosomes may be either non-random or random depending on the growth conditions. Cell division is closely linked to chromosome replication and it is postulated that the prokaryotic cell cycle consists of a DNA-division cycle and a growth cycle. The process of cell septation is at present poorly understood (see James, 1978).

Eukaryotic DNA-division cycle

In this section we will consider the general features of the eukaryotic DNA-division cycle and indicate some of the differences which may be encountered in eukaryotic micro-organisms. It is unrealistic to attempt a comprehensive account here and we will deal only with the main processes and how they compare with similar events in prokaryotes.

Chromosome replication The DNA of eukaryotes is organized into a number of chromosomes which are located in a membrane-bound nucleus distinct from the cytoplasm. The exact number of chromosomes per cell is species-dependent. A nucleolus is also present, and the DNA is closely associated with specific basic proteins called *histones*. The phases undergone by DNA during the cell cycle have been called G1, S, G2 and mitosis, followed by cell division. The G1 period is a phase of non-synthesis of DNA during which many other important events occur whose function is to prepare for the S or synthetic period. It is during the S-period that chromosome replication, and hence DNA synthesis, occurs. The synthetic period is followed by the G2 phase, which is also a non-synthetic period and leads into mitosis, followed by cell division. A further resting phase called the G0 phase also occurs in which cells remove themselves from their cell cycles into a resting phase. The G0 phase is usually found in higher eukaryotic cells.

The DNA-division cycle

In plant or animal cells the duration of each of the G1, S, G2 and mitotic phases is fairly constant. Nonetheless, certain factors must be borne in mind when considering chromosome replication of eukaryotes, especially in eukaryotic micro-organisms such as yeasts, amoebae, flagellates, ciliates, and unicellular algae which differ from plant and animal cells in the distribution and relative amounts of their DNA.

First, some of the DNA is localized outside the nucleus in organelles such as mitochondria or chloroplasts. In some eukaryotic micro-organisms up to 20 per cent of the total DNA content of the cell may be cytoplasmic (in other words, present in organelles other than the nucleus). The interaction between the synthetic periods of nuclear and cytoplasmic DNA formation will be dealt with later.

Second, some protozoa such as *Tetrahymena* and *Paramecium* contain one or more micronuclei in addition to the macronucleus. The phases undergone by the micronuclear DNA are always different from those of the nucleus.

Third, unlike DNA synthesis in prokaryotes, DNA synthesis in eukaryotes is *always periodic* (although amphibian embryos have been reported to carry out continuous DNA synthesis).

Finally, different eukaryotic micro-organisms differ markedly in the proportion of the cell cycle spent in the different phases. Certain protozoa, such as *T. pyriformis* and *Amoeba proteus*, have no detectable G1 period while others lack a G2 period. Various yeasts differ markedly in the relative portions of the G1, S and G2 phases. *S. cerevisiae* has been reported to have G1, S and G2 periods of approximately equal lengths, while *S. pombe* has no G1, a very short S-phase of approximately 10 min and then a G2 period of 2.5 hours. In the unicellular alga *Chlorella pyrenoidosa*, the S-period occurs towards the end of the cycle. This organism therefore has a long G1 and a short G2 period.

The DNA of eukaryotes is arranged in linear chromosomes each of which is organized into discrete regions which have been termed *nucleosomes*. Each nucleosome consists of some 150–200 base pairs packaged in nucleosome sub-units, the size and conformation of which are determined by the chromosomal proteins, in particular the histone proteins. These discrete nucleosome regions are linked to each other by inter-particle DNA regions which can vary in numbers of base pairs from a few to approximately 150.

DNA replication unit The concept of a unit of DNA replication initially stemmed from studies of the replicon of prokaryotes. This term has now been expanded to incorporate the molecular and genetic events which constitute the replication of the single DNA element of bacteria and their phages, the simple genomes of viruses,

The microbial cell cycle

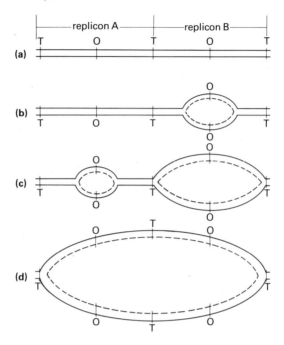

Fig. 3.8 Model for DNA replication in eukaryotes.
a The two horizontal lines represent a section of double-stranded DNA comprising two replicons A and B, each of which has an origin (O) and two termini (one of which is shared).
b Replicon B initiates replication first. Newly synthesized DNA is represented by dotted lines.
c Replicon A then initiates replication, by which time the two replication forks in B have reached their respective termini.
d Replicon A finishes replication and sister DNA helices separate. (Redrawn from Huberman & Riggs, 1968.)

and the basic replicating unit of the eukaryotic chromosome. It is proposed that the linear eukaryotic chromosome is made up of a large number of replicons, each of which consists of an origin and two termini of replication. *Replicons are not replicated sequentially* from one end of the chromosome to the other. A complex order of replication exists so that certain areas of the chromosome are replicated early in the S-phase while others replicate at later stages of the S-phase. This apparently random process is repeated in successive cycles; that is to say, those replicons which replicate early in the S-phase of one cycle will do so in successive cycles. This rather haphazard order of replication was observed by adding a pulse of tritiated thymidine to growing cells and then examining them by autoradiography at intervals after the pulse. It was noted that the labelling of chromosomes was not uniform, implying that different regions replicated at different times during the S-phase. A model summarizing the main features of eukaryotic chromosome replication is shown in Fig. 3.8.

Replication is initiated at the origin of a replicon and proceeds bidirectionally *via* two replication forks to two different termini. A single terminus can terminate two different replicons. Regions of the same chromosome replicate at different times of S-phase. The rate of chromosome replication of eukaryotes is approximately $1 \mu m \, min^{-1}$, some ten times slower than in *E. coli*. The control of initiation of chromosome replication and the elements responsible for it are not known. Whatever the mechanism, it must explain the time-ordered synthesis of different regions of the chromosomes. This process indicates that replication is subject to a high degree of cellular control. The signal may well be an initiator protein because DNA synthesis can be blocked by inhibitors of protein synthesis, especially towards the beginning of S-phase. Evidence that the initiator protein is cytoplasmic comes from the synchronous mitoses which occur in multi-nucleate cells. These are a common feature of the lower eukaryotes such as *P. polycephalum* and *A. castellanii*.

Perhaps the most startling evidence for a cytoplasmic initiator comes from studies of nuclear transplants. Nuclei from macrophages which normally do not divide will do so if they are transplanted into the cytoplasm of another cell type. The resulting binucleate cell undergoes a synchronous S-phase of the macrophage nucleus and its own nucleus.

The S-phase As in prokaryotes, once DNA synthesis is initiated, a train of events is set in motion resulting in the duplication of DNA by a semi-conservative mechanism. The elements responsible for replication are not as well characterized as in prokaryotes, but include DNA polymerase, DNA ligating enzymes, single-stranded DNA-dependent ATPases, DNA binding proteins, endo and exo-DNAses, RNA polymerase, histones, and other chromosomal proteins.

The rate of replication—that is, movement of the replicating complex—is very similar in most eukaryotic micro-organisms, ranging from 0.1 to $2 \mu m \, min^{-1}$ (corrected to 37°C), which is much slower than in *E. coli*. The rate of progress of a replicating fork may well be dependent upon genetic and temporal controls as well as on the prevailing growth conditions. In yeasts and *P. polycephalum*, 80 to 90 per cent of nuclear DNA is replicated in the first 10 to 25 per cent of S-phase. However, it seems that, as in prokaryotes, control of the rate of replication will lie at the site of initiation and any possible alteration of the actual rate is very small and acts chiefly as a fine control of the S-phase. Little is known about the specific nucleotide sequences which may code for an origin or terminus of replication. The DNA helix consists of two strands of opposite polarity and the DNA chain elongation enzymes are of single direction specificity. It was therefore originally proposed that since DNA was polymerized in both directions at the same time the synthesis of one strand was discontinuous. However, recent evidence from a number of different cell types ranging from eukaryotic micro-organisms to plant and animal cells suggests that DNA synthesis may be discontinuous in both strands. It is probable that protein synthesis is necessary throughout the S-phase. In contrast, certain yeasts can complete chromosome replication in the absence of protein synthesis once initiation has occurred.

An important event during the S-phase is the synthesis of the histone proteins at the same time as that of DNA. It appears that this tight coupling between histone and DNA synthesis is found in most eukaryotic cells. It is thought that newly synthesized DNA becomes associated immediately with histone proteins, of which there are many classes. They may, therefore, prevent the further initiation of chromosome replication in newly synthesized DNA and act in a similar fashion to the hypothetical repressor protein involved in initiation of chromosome replication in prokaryotes. Histones have also been assigned a structural role in eukaryotic chromosomes. Their close association with the DNA is demonstrated by the observation that inhibitors of DNA synthesis, such as high levels of thymidine or hydroxyurea, also block histone synthesis. Since histones (and other proteins) make up ordered and tight complexes with DNA, the replicating complex must be able to dislodge or temporarily disrupt them in order to replicate the genome. This may account for the much slower rate of chromosome replication observed in eukaryotic cells.

Replication of extrachromosomal DNA We have already noted that certain ciliated protozoa such as *Paramecium aurelia* contain DNA localized in micronuclei. The phases undergone by this DNA differ markedly from those of the macronucleus, especially the S-period which usually lasts for a shorter time. Eukaryotic cells, including eukaryotic micro-organisms, also have DNA units located in the nucleolus. Nucleolar DNA from *T. pyriformis* and *P. polycephalum* has been isolated and characterized. Many of the sub-units code for ribosomal RNA (rRNA) *via* palindromic base sequences (sequences which are the same when read from the 3' end to the 5' end from either DNA strand). Isolation of nucleolar DNA which encodes rRNA from *Tetrahymena* has revealed that replication is bidirectional from the centre of the linear palindromic molecules. The *Physarum* DNA encoding rRNA is synthesized throughout the inter-mitotic period with the exception of the first hour of the S-period. If prolonged mitotic cycles are artificially induced in this organism, an over-replication of nucleolar DNA occurs. Although during normal cell cycles the amount of DNA encoding rRNA doubles, not all of the units are replicated, indicating that some are replicated more than once.

Eukaryotic cells also contain extra-chromosomal DNA located in mitochondria and in chloroplasts of photosynthetic cells. In plant and animal cells the amount of organellar DNA is usually small, amounting to about 1 per cent of total cellular DNA. In eukaryotic micro-organisms such as yeasts, protozoa, and algae, organellar DNA may constitute 15 to 20 per cent of total cell DNA. This is typified by members of the Trypanosomatidae (a family of parasitic protozoa), members of which pack 20 per cent of their cell DNA into a discrete region of the single mitochondrion called the *kinetoplast*. The kinetoplast DNA of the genera *Crithidia* and *Leishmania* consists mainly of approximately 27 000 covalently closed circular DNA molecules, some of which are catenated (interlocked into a chain). It appears that all the circular DNA molecules, which are approximately 0.8μ long, carry the same genetic information. During replication of kinetoplast DNA, long linear molecules become apparent.

Eukaryotic micro-organisms have a degree of autonomy in the biogenesis of their mitochondria, and this we shall discuss later.

Control of cell division The controls responsible for the integrated completion of cell cycle events preceding cell division are not well understood. Many approaches have been adopted in an attempt to understand the timing of cell division in relation to other parameters such as nuclear division. Perhaps the most productive system for studying this problem has been that of heat-synchronized cultures of *T. pyriformis*. This protozoon was one of the first organisms used for studying aspects of the cell cycle, synchrony being achieved by the use of heat shocks. This method for establishing synchronous cultures was described on page 13. Attempts to explain how heat shocks caused synchronous cell division gave rise to the formulation of a model in which it was proposed that different stages of the cell cycle are differentially sensitive to the heat shocks and this has the effect of delaying division. Cells early in the cycle will be relatively unaffected and will divide more or less at the correct time. Cells later in their cycles will be progressively affected to a greater extent, and the delay before division will be correspondingly lengthened. This increased

The microbial cell cycle

Table 3.1 Summary of DNA replication in prokaryotes and eukaryotes

Prokaryotes	Eukaryotes
DNA is probably organized into a single circular chromosome lying free in the cytoplasm apart from one or more points of attachment to the cytoplasmic membrane.	DNA is organized into a number of linear chromosomes which are bounded by a nuclear membrane.
Extrachromosomal DNA present as plasmid DNA.	Extrachromosomal DNA is present in the nucleolus, mitochondria and chloroplasts, and in micronuclei of ciliates.
Phases undergone by DNA in slowly growing bacteria can be divided into I, C and D-periods, followed by division.	Phases undergone by DNA are divided into G1, S, G2, mitosis, and division.
DNA is replicated during the C-period and may be periodic during the cell cycle of slowly growing cells or continuous during that of rapidly growing cells.	DNA synthesis occurs during the S-period and is always periodic.
Bacteria can initiate chromosome replication before the previous round has been completed by means of multiple replication forks.	Chromosome replication occurs once every cycle, but a single chromosome contains many replicating units.
Replication begins at a single origin, proceeds bidirectionally, and ends at a common terminus.	Replication can occur at many origins, proceeds bidirectionally, and ends at two termini per origin.
Replication rate in the order of $14 \mu m\,min^{-1}$ for *E. coli*.	Replication rate in the range $0.1-2 \mu m\,min^{-1}$.
Chromosome segregation probably depends on the growth of the membrane and/or generation of new attachment sites.	Chromosome segregation is achieved by means of a complex mitotic apparatus.
Division dependent on gene products synthesized early on in the cycle, possibly at initiation.	Possibly the same as in prokaryotes.
'Histone-like' proteins occur in *E. coli* but their role is as yet unclear.	Histones play an important part in DNA replication and maintenance of chromosome structure.

delay in division in cells at later stages of their cycle stops at a critical point late on in the cycle before division. It has been termed the *transition point*. Cells at or after the transition point are committed to divide and will be unaffected by heat shocks.

It is postulated that other periodic induction methods (starvation or amino acid deprivation, for example) induce synchronous growth by a similar differential effect at different stages of the cell cycle. One proposal to explain these observations is that division proteins are synthesized throughout the cell cycle. They are necessary for division up to the transition point and are postulated to be extremely unstable up to this point. During the heat shocks they are broken down completely but it is supposed that at the transition point they are in some way stabilized so that afterwards they are able to withstand the heat shocks. This model may have wide applications in explaining the effects of heat shocks used to prepare synchronous cultures of bacteria such as *E. coli* as well as other eukaryotes.

Summary

A comparison of the main events occurring during DNA replication of prokaryotes and eukaryotes is presented in Table 3.1.

References

COOPER, S., SCHWIMMER, M. & SCANLON, S. (1978). Probabilistic behaviour of DNA segregation in *Escherichia coli*. *Journal of Bacteriology* 134: 60–65.

FORTERRE, P., RICHET, E. & KOHIYAMA, M. (1978). The role of ATP in *E. coli* chromosome replication. In: *DNA synthesis: present and future* pp.397–407. Edited by I. Molineux and M. Kohiyama. Plenum, New York.

HELMSTETTER, C.E. (1969). Regulation of chromosome replication and cell division in *Escherichia coli*. In: *The Cell-Cycle, Gene-Enzyme Interactions* pp.15–35. Edited by G.M. Padilla. G.L. Whitson, and I.L. Cameron.

HUBERMAN, J.A. & RIGGS, A.D. (1968). On the mechanism of DNA replication in mammalian chromosomes. *Journal of Molecular Biology* 32: 327–41.

JACOB, F., BRENNER, S. & CUZIN, F. (1963). On the regulation of DNA replication in bacteria. *Cold Spring Harbour Symposium of Quantitative Biology* 28: 239–48.

JAMES, R. (1978). Control of DNA synthesis and cell division in bacteria. In: *Companion to Microbiology* ch.3. Edited by A.T. Bull and P.M. Meadow. Longman, London.

KLEPPE, K., ÖVREBÖ, S. & LOSSIUS, I. (1979). The bacterial nucleoid. *Journal of General Microbiology 112*: 1–13.

PRITCHARD, R.M. (1978). Control of DNA replication in bacteria. In: *DNA synthesis: present and future* pp.1–26. Edited by I. Molineux and M. Kohiyama. Plenum, New York.

SARGENT, M.G. (1974). Nuclear segregation in *Bacillus subtilis*. *Nature* 250: 252–254.

SARGENT, M.G. (1978). Surface extension and the cell cycle in prokaryotes. *Advances in Microbial Physiology* 18: 105–176.

4 RNA and protein synthesis

During the DNA-division cycle the cell must grow and generate sufficient cell mass to produce two complete daughter cells at division. This growth cycle consists of a multitude of cellular reactions which convert external nutrients into internal cellular products. The reactions will involve the catabolic breakdown of nutrients to conserve energy in the form of ATP and reducing power as reduced pyridine nucleotides (NADH and NADPH), some of which will be used in a series of anabolic reactions to produce new cell material.

The growth cycle will be affected by temperature, nutrient availability, pH, and so on. During balanced growth, the cycle will be controlled largely by transcription/translation reactions and their products. The most important of these products will be enzymes and structural proteins. Control of protein synthesis during the cell cycle is, therefore, of paramount importance. Sufficient amounts of a particular protein must be present at the correct stage of the cycle and in the correct region of the cell. For periodically synthesized proteins both temporal and spatial controls must operate. Little is known of how transcription and translation reactions are coupled during the cell cycle. Many components of the processes have been characterized, but it is difficult to investigate their expression during the cell cycle especially since some—certain types of mRNA, for example—are discontinuous in their activity.

The bulk of our knowledge about how transcription and translation reactions are regulated has come from studies of the protein end-products. Two general approaches have been adopted. One is to assay activities of individual enzymes during the cell cycle; the other is to measure amounts of non-enzymic protein during the cycle. Perhaps the best studied of the latter type of proteins are the histones which are associated with nuclear DNA of eukaryotes.

In this chapter we will concentrate on RNA and enzyme synthesis during the cell cycle, how they differ in prokaryotes and eukaryotes, and draw attention to some of the obstacles and difficulties which are encountered in their determination. Possible control mechanisms which may operate during enzyme synthesis will also be discussed.

RNA and protein synthesis

RNA synthesis

The process of transcription and the participating elements have recently been reviewed by Burdon (1976). Certain genes code for transfer RNA (tRNA) and others for ribosomal RNA (rRNA): once synthesized, the RNA molecules become extensively folded. The rRNA becomes associated with specific proteins to form ribosomes. Transcription of other specific sequences of the genome results in the

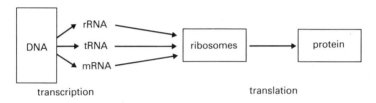

Fig. 4.1 Coupling of transcription with translation.

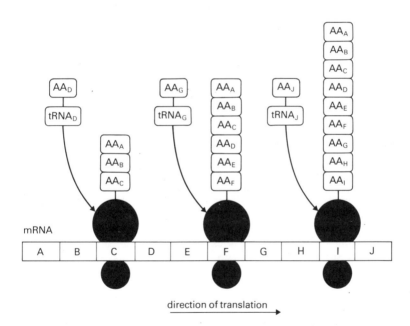

Fig. 4.2 Protein synthesis at a polysome of three ribosomes with a mRNA molecule. Each of the boxed letters A to J refers to a codon on the mRNA molecule. As the ribosomes move along the molecule the amino acids are added sequentially by the corresponding tRNA molecules to the growing polypeptide chain. The greater the number of ribosomes associated with the mRNA, the more protein will be synthesized.

biosynthesis of a third RNA class, messenger RNA (mRNA), which will eventually be translated at the ribosomes to synthesize proteins with the participation of tRNA and amino acids. This process is summarized simply in Fig. 4.1.

The RNA species which was characterized first was tRNA; it comprises 10 to 15 per cent of total cell RNA in both prokaryotes and eukaryotes. The bulk of the cellular RNA pool, 80 per cent, exists as rRNA, and the remaining 3 to 5 per cent is mRNA. Each molecule of mRNA codes for the synthesis of a specific protein chain. It is transcribed on the chromosome and then moves to the ribosomes with which it becomes associated to form a chain of ribosomes called a *polysome* (Fig. 4.2). Each ribosome moves along the mRNA molecule and translates the information into protein. The more ribosomes that become attached to an mRNA molecule, the more protein will be synthesized. Some mRNA molecules code for more than one protein; such messenger molecules are termed *polycistronic*.

RNA synthesis in eukaryotic micro-organisms The bulk of the cellular RNA of eukaryotes is cytoplasmic and only about 5 per cent of the total is found in the nucleus. It is now thought that the RNA of the nucleus is in a continuous state of flux and comprises a 'mini-pool' of several species of RNA which are intermediates in biosynthetic pathways leading to the formation of the different RNA classes. The primary structure of these RNA intermediates can be modified in certain instances by cellular enzyme systems. Most of the nuclear RNA is contained in the nucleolus, which is extremely active, giving rise to rRNA.

There are formidable problems in following RNA synthesis during the cell cycle, some of which will apply equally to prokaryotes and eukaryotes. These include:

> Difficulties in determining the concentrations of the three RNA classes during the cell cycle. Each exists in different proportions, and RNA molecules are subject to *post-transcriptional* processing by 'cutting' enzymes. Some of these alter the primary nucleotide sequences by modifying bases; others add on specific nucleotide sequences.

> Eukaryotic micro-organisms have organellar DNA and RNA located in chloroplasts and/or mitochondria. Mitochondrial ribosomes differ from those of the cytoplasm in their sedimentation properties (sedimentation coefficient 70s instead of 80s) and probably in their RNA composition.

Measurements of RNA levels during the cell cycle have either involved determining the total cell RNA or measuring the rate of

RNA and protein synthesis

RNA synthesis at different stages of the cycle. Generally, total RNA is synthesized continuously throughout the cell cycle as in *C. fasciculata* and *S. pombe*, so that the total quantity doubles every cycle. Determinations of the rates of RNA synthesis through the cycle have been attempted using radioactively labelled RNA precursors. A major obstacle here is the different pool sizes of the various RNA classes. One class may be labelled at a different rate from another. Recently, the rate of polyadenylated messenger RNA synthesis has been investigated during the cell cycle of *S. pombe* using wild-type cells, and a mutant (wee 1–50) in which mitosis and cell division occur in cells about one half the size of the wild-type (Fraser & Nurse, 1978). From this work it appears that a size-related control mechanism operates which maintains an average mRNA content in balance with total cell mass during growth, even in cells dividing at different sizes and growing at different absolute growth rates per cell. It is proposed that this size-related mechanism for regulating the time at which the rate of mRNA synthesis doubles may regulate the synthesis of those enzymes not subject to specific controls at the transcriptional and translational levels. We will deal with some of these specific controls later.

RNA synthesis in prokaryotic micro-organisms Little information is available about the synthesis of different RNA classes during the cell cycle of prokaryotes. It appears that the total rate of RNA synthesis is closely coupled to the synthesis of other macromolecules. In *E. coli* this coupling may be controlled by the rate of RNA polymerase synthesis. The amount of RNA synthesized by a bacterium is limited by the rate of initiation of an RNA chain. In other words, once the synthesis of an RNA chain has been initiated, RNA polymerase moves away from the promoter site and transcribes the genetic material. The initiation site is now available to a second RNA polymerase molecule. The frequency of initiation will depend upon the number of RNA polymerase molecules the fraction of the genome is able to accommodate at any one time, and which is available to initiate synthesis. Important factors in regulating the rate of RNA synthesis during the cell cycle therefore include the regulation of the rate of synthesis and structural modification of RNA polymerase.

Summary

Under standard conditions, the growth of an organism during the cell cycle is controlled by transcription and translation reactions. These have important temporal and spatial controls. Measurement of the rate of RNA synthesis during the cell cycle is complicated by the presence of three RNA classes, each of which has a different pool size. Little is known about the relative patterns of synthesis of the different types of RNA in either eukaryotes or prokaryotes but it seems that *total* cellular RNA is synthesized continuously during the cell cycle.

Protein synthesis

The total complement of cell protein includes soluble protein as well as protein in close association with other molecules such as lipoproteins and glycoproteins and macromolecules such as RNA (in ribosomes) and DNA. Proteins usually fulfil enzymic or structural roles. Measurements have revealed that the total cell protein complement increases continuously throughout the cell cycle in parallel with RNA synthesis and increase in dry weight. It is only when one begins to look at patterns of synthesis of individual proteins that one realises that there is a temporal complexity of protein synthesis in both prokaryotes and eukaryotes. This complexity is masked by many of the methods used to determine the rate of synthesis or amount of protein—for example, estimation of rates of incorporation of labelled amino acids into the protein pool.

The most widely used approach to investigate the expression of individual proteins during the cell cycle is to assay enzyme activities in synchronously growing cells. This is an easy and accurate method for most enzymes which has provided valuable information concerning cellular processes such as metabolic regulation and control of gene expression during the cell cycle (Halvorson *et al.*, 1971). Before embarking on the state of current knowledge of enzyme synthesis during the microbial cell cycle it is wise to list some cautionary points which apply equally to prokaryotes and eukaryotes:

Observed enzyme activities *in vitro* do not necessarily reflect the amount of enzyme protein *in vivo*.

When assaying enzyme activities, all substrates must be in sufficient excess to allow for any change in the Km of an enzyme for its substrate during the cell cycle.

When determining specific activity of an enzyme during the cell cycle no loss or leakage of protein must occur during the extraction procedure otherwise abnormal specific activities may be noted.

During removal of samples and assaying, care must be exercised to ensure that no co-factor essential for activity is lost or diluted.

Enzyme activity *in vitro* may not always reflect activity *in vivo* since one can recognize two levels of enzyme regulation. These

RNA and protein synthesis

are the *internal regulation* imposed by cellular processes and, more importantly when considering enzymes from synchronous cultures, the *external regulation* imposed by the cell's environment. The latter will promote long-lasting effects if the method used to initiate synchronous growth is an especially perturbing one.

Patterns of enzyme synthesis during the cell cycle Many enzymes have been assayed through the cell cycle, mostly in cell extracts prepared at different stages of synchronous growth. Measurements of 'marker' enzymes (enzymes associated with a definite subcellular location or process) can lead to an understanding of spatial as well as temporal controls. Good examples here include those enzymes associated specifically with a subcellular organelle in the eukaryotic cell such as acid phosphatase in lysosomes or cytochrome *c* oxidase of the inner mitochondrial membrane, or processes such as bacterial cell wall synthesis.

From a wide range of investigations on diverse species of prokaryotic and eukaryotic micro-organisms we now know that the pattern of enzyme synthesis during the cell cycle may be *periodic* or *continuous*. Continuously synthesized enzymes may increase exponentially or linearly during the cell cycle (Fig. 4.3). In contrast, the amount of an enzyme synthesized periodically doubles at a temporally fixed point in the cell cycle. This may occur in a step-wise fashion for *stable enzymes* (those whose activity remains constant), or as a series of maxima and minima for *unstable enzymes* (those whose activity is subject to decay). (Fig. 4.3). Certain periodically synthesized enzymes can have more than one period of synthesis per cell cycle, so that activity increases as a series of steps or maxima. The overall effect under conditions of balanced growth will be to double the enzyme amount for the ensuing division. An example of an enzyme which is synthesized continuously is ß-galactosidase in selection-synchronized *E. coli*. Malate dehydrogenase, on the other hand, is synthesized periodically in selection-synchronized cultures of *S. cerevisiae*. There appears to be no universal trend in the patterns of enzyme synthesis. Periodic synthesis of an enzyme in one organism may be replaced by continuous synthesis of the same enzyme in another organism.

Repression and de-repression of enzyme synthesis In both eukaryotes and prokaryotes one is faced with the problem that the observed enzyme activity may not be the same as the cell's maximum possible output. Enzyme activities are highly dependent on the nature of the growth medium. Certain chemicals repress the rate of synthesis of some enzymes while enhancing that of others. A good example is

The microbial cell cycle

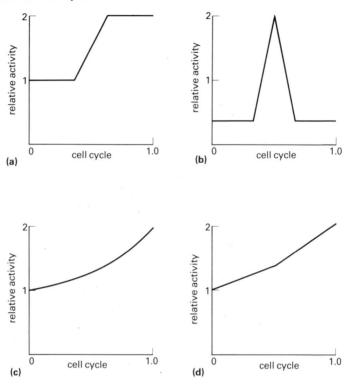

Fig. 4.3 Possible patterns of enzyme activity observed during a single cell cycle, expressed as an arbitrary period 0 to 1.0. Periodic activity is expressed as a step increase (a) or as an unstable peak increase (b). Continuous synthesis is observed as an exponential increase (c) or a linear increase (d). (Redrawn from Mitchison, 1971.)

afforded by *S. pombe*, in which many of the enzymes of terminal respiration are repressed in cells growing synchronously in the presence of 1% (w/v) glucose. Many of these enzymes are expressed as unstable 'peak' enzymes during the cell cycle (Fig. 4.3), or as a stable 'step' pattern of increase in cells growing synchronously with glycerol under conditions of de-repression (Edwards & Lloyd, 1977). Another problem is presented by inducible enzymes which, in the absence of the appropriate inducer, will be present at basal levels. These may amount to only a few enzyme molecules per cell, whereas the actual synthetic capacity of the cell at a given time may be many hundreds of molecules per cell.

An approach that permits determination of enzyme repression and the cell's capacity for synthesis through the cycle is to grow cells in synchronous culture in the *absence of inducer*. Samples are removed during synchronous growth, put into a medium containing the inducer, and the rate of enzyme synthesis is then assessed. From such experiments two major classes of inducible enzymes are apparent: those in which enzyme synthesis is maximal after addition of the inducer *at all stages of the cell cycle*, and those in which enzyme synthesis is maximal after addition of the inducer *only at certain fixed stages of the cell cycle*.

The first mechanism is commonly encountered in prokaryotes and is exemplified by the induction of ß-galactosidase by thiomethyl galactoside in asynchronous cells of *E. coli*. Within a few minutes of adding the inducer *all* the cells produce the enzyme at their maximal rates. This implies that enzyme synthesis occurs at any stage of the cell cycle. The second mechanism is generally detected in eukaryotic micro-organisms. For example, in the alga *Chlorella pyrenoidosa*, the enzyme nitrite reductase can be induced by nitrite only when DNA synthesis has commenced. The induction is sensitive both to actinomycin D and chloramphenicol. There are exceptions: the enzyme sucrase is fully inducible at all stages of the cell cycle of *S. pombe*.

Summary

Total cell protein increases continuously during the cell cycle in concert with increase in cell mass and total cellular RNA. Synthesis of individual proteins may be continuous or periodic. Assay of enzyme activities during the cell cycle is complicated by a variety of factors, especially repression and de-repression by growth factors. Inducible enzymes may be either fully inducible throughout the cell cycle, a situation found in most prokaryotes, or only at certain temporally fixed stages of the cycle, a situation more common to eukaryotic micro-organisms.

Models of microbial enzyme synthesis during the cell cycle

Prokaryotes growing under standard conditions will synthesize some enzymes at basal or repressed levels and others at maximum levels in response to inducers in the growth medium. Enzymes will be maintained at intermediate levels in response to the cell's internal controls. Generally, biosynthetic enzymes will be subject to end-product repression, while degradative enzymes will be subject to catabolite repression. In the fully repressed state, an enzyme may exist at an extremely low copy number. For example, ß-galactosidase is synthesized continuously by *E. coli* throughout the cell cycle, maintaining a basal level of a few molecules per cell. In the presence of an inducer, enzyme levels increase dramatically at any stage of the cell cycle. Enzymes subject to strict internal controls will be synthesized at definite times during the cycle, resulting in a periodic peak, step, or rate increase every cell cycle.

Most of the information about enzyme synthesis during the cell cycle of eukaryotic micro-organisms comes from studies with yeasts. The difference between fully repressed and fully induced enzyme levels is not as great in eukaryotic cells as it is in prokaryotes. In the former, the enzymes may be induced about one hundred-fold whereas, in bacteria, induction factors of a thousand-fold or more are observed. This is probably due to the much more complex internal architecture of the eukaryotic cell, in which enzymes are partitioned into organelles.

Much of our understanding of the controls operating on inducible enzymes have stemmed from investigations into the induction of lactose-metabolizing enzymes in *E. coli*. These investigations led to the hypothesis of the *lac operon* with its control and regulation of enzyme synthesis (Fig. 4.4). A brief outline of this hypothesis is presented here since it raises a number of interesting points which may apply to control of enzyme synthesis during the cell cycle. The catabolism of lactose by *E. coli* requires three enzymes, ß-galactoside permease which allows lactose to enter and become concentrated within the cell; ß-galactosidase which hydrolyses lactose to glucose and galactose; and thiogalactoside transacetylase, whose function is not properly understood. The main features for the regulation of expression of the *lac* operon have been summarized by Walker (1977):

> Enzyme synthesis is regulated by means of information transmitted to the genome from the cytoplasm.

> The enzymes are normally maintained at basal levels because a repressor protein prevents transcription of the operon; the inducer overcomes this control.

> The repressor molecule is produced at a constant rate of 1 to 2 molecules per cell and prevents transcription by attaching itself to a specific region of the operon known as the *operator*. The concentration of the repressor is unaffected by the inducer.

> Transcription and translation are tightly coupled so that the rate of enzyme synthesis is controlled largely by the rate of transcription.

It is apparent from investigations of the *lac operon* that regulation of enzyme synthesis can be mediated *via* a repressor protein synthesized constitutively by a regulatory gene at a site separate from the operon. Another more recently discovered operon exhibits a

RNA and protein synthesis

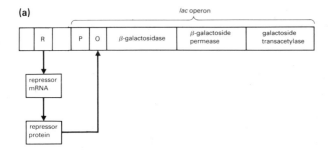

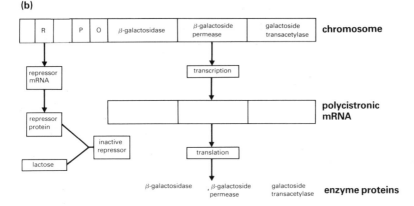

Fig. 4.4 The lac operon.
a In the absence of inducer (lactose) the constitutively produced repressor protein binds to the operator and prevents transcription.
b The inducer inactivates the repressor protein and allows RNA polymerase to transcribe the genome. Transcription gives rise to the production of polycistronic mRNA which, at translation, results in the co-ordinated production of the three enzymes.

different type of regulation. The repressor protein is coded on a gene which is part of the operon itself. The repressor, therefore, controls the synthesis of its own mRNA. This type of 'autogenous regulation' is operative in histidine catabolism in *S. typhimurium*.

The ability of bacterial populations to respond within a few minutes to changes in their environment has important ramifications when considering observed patterns of enzyme synthesis during the cell cycle. Models such as the *lac operon* have led to the concept of the *oscillatory repression model* or *feedback control of enzyme synthesis*. This hypothesis has been developed chiefly to explain enzyme synthesis during the prokaryotic cell cycle.

The microbial cell cycle

Oscillatory repression model This model proposes that the end-products of enzyme reactions repress the synthesis of the enzyme by *negative feedback* which, under constant conditions, leads to oscillations in the levels of enzyme activity during the cell cycle. There is a lowering of enzyme activity when products accumulate and synthesis of more enzyme when concentrations of end-products are low. However, enzyme levels in synchronous cultures oscillate with a definite periodicity. They rise to a maximum and fall to a minimum at the same stage of the cycle in successive divisions. To explain this, it has been proposed that the entrainment of the oscillations is ordered by a strict temporal event such as a pulse of mRNA synthesis at gene replication.

Certain predictions can be made from the model and these can be tested experimentally. Under some conditions, the rate of enzyme synthesis should be fully induced or de-repressed throughout the cell cycle. In *E.coli*, fully induced and constitutive ß-galactosidase activities increase continuously during the cell cycle, in keeping with this prediction. However, the situation is complicated by the observation that fully induced sucrase activity in synchronous cultures of *B. subtilis* is periodic (Donachie & Masters, 1969).

The synthesis of autogenously regulated enzymes should be periodic during the cell cycle. That is, enzyme levels should rise in response to diminishing concentrations of end-products, and *vice versa*. Evidence for this is not so readily obtained because enzyme levels will depend on the *rate of formation* of the end-product as well as its *rate of utilization*. Each will vary according to the prevailing conditions. A large number of enzymes exhibit periodic oscillations during the bacterial cell cycle but this is not a common occurrence in eukaryotes. The alga *Chlorella*, for example, synthesizes the enzyme ribulose 1,5 diphosphate carboxylase continuously at high growth rates, when synthesis is presumably de-repressed, and periodically in slowly growing cells, when the enzyme is repressed.

The potential for the maximum possible amount of enzyme synthesis of a cell should exist at all times during the cycle and should reflect gene dosage. For example, an inducible enzyme, normally repressed in the absence of inducer, should be synthesized maximally at any stage of the cycle on addition of the appropriate inducer. In a synchronous culture cells should be able to induce the maximum rate of enzyme synthesis at any time and double the induction capacity at a point in the cycle corresponding to that at which the gene coding for that enzyme is replicated. Evidence for both requirements has been obtained from synthesis of the inducible lactose-utilizing enzymes during the cell cycle of *E. coli*.

Summary

Much of our understanding of the regulation of enzyme synthesis in prokaryotes has come from work on the *lac* operon of *E. coli*. This has led to the formulation of the model of *oscillatory repression* of enzyme synthesis during the prokaryotic cell cycle, especially for inducible enzymes. Experimental evidence indicates that the genome of prokaryotes is available for transcription and translation at all stages of the cell cycle.

Many observations made with eukaryotes are inconsistent with the oscillatory repression model for the regulation of enzyme synthesis. This is probably due to the greater structural organization and the likely additional control mechanisms found in these organisms. However, a model has been proposed to explain regulation of enzyme synthesis during the eukaryotic cell cycle. This is referred to as the *sequential transcription model* or, more often, the *linear reading model*.

Linear reading model According to this model, the ordered temporal appearance of certain enzymes during the cell cycle of many eukaryotic micro-organisms reflects the sequential transcription of the genes coding for those enzymes. It is postulated that there is a linear relationship between the *time* of enzyme appearance and the *position* of the genes on the chromosome. Genes are transcribed in sequence as RNA polymerase moves along the chromosome. The model predicts, therefore, that:

> A gene is available for transcription and, therefore, for induction for only a short time during the cycle.

> There is a definite unique period in the cell cycle at which a particular gene is transcribed.

> Periodic synthesis of enzymes should occur throughout the cell cycle.

Much of the supporting evidence for this model comes from studies of enzyme synthesis during the yeast cell cycle. There is plenty of evidence to show that enzyme synthesis occurs throughout the cycle and that inducible enzymes respond to inducers only during a restricted period of the cycle (Halvorson *et al*, 1971). Evidence for the second prediction is provided by studies with a hybrid of two yeasts, *Saccharomyces dobzhanskii* and *Saccharomyces fragilis*. The enzyme ß-glucosidase of each yeast was synthesized periodically in a stable, single-step pattern during the cycle, but the hybrid produced *two* step increases of ß-glucosidase activity. Furthermore, the two enzymes were antigenically distinguishable, so it was possible to demonstrate that there were unique periods of transcription for the two genes coding for ß-glucosidase in the hybrid. If oscillatory repression had been operating, only *one* periodic step increase of activity would be expected.

The model is not applicable to synthesis of all enzymes in eukaryotes, especially those synthesized continuously during the cell cycle of *S. pombe* such as acid phosphatase, sucrase and alkaline phosphatase. Another group of enzymes which do not conform to the model are those which remain fully inducible at all stages of the cycle. Clearly, other controls must also regulate the synthesis of enzymes during eukaryotic cell cycles. These may include translational control of long-lived mRNA molecules.

Summary

The linear reading model has been proposed to explain the observed patterns of enzyme synthesis during the eukaryotic cell cycle. Other control mechanisms must exist, however, since the model does not cover all experimentally observed patterns of enzyme synthesis.

Other control mechanisms

Some other possible mechanisms for regulating the synthesis of enzymes during the cell cycle are presented in Fig. 4.5. These are applicable to both prokaryotes and eukaryotes.

Prokaryotes and eukaryotes also have internal structural organization which will affect controls on enzyme synthesis. These are summarized in Table 4.1.

Regulation of enzyme synthesis in eukaryotes is further complicated by the presence of a limiting nuclear membrane. The possible control mechanisms within the eukaryotic nucleus are shown in Fig. 4.6. Once translation has been initiated, the enzyme molecules produced are under the controls already outlined.

Discontinuous genes Recently it has become apparent that the eukaryotic chromosome may have blank areas—that is, DNA sequences coding for proteins are separated by regions with unknown functions (Flavell *et al.*, 1978). These observations stem from experiments involving the cloning of eukaryotic DNA by recombining it with bacterial plasmids or bacteriophage DNA *in vitro*. These regions could be involved in unknown control mechanisms for transcription.

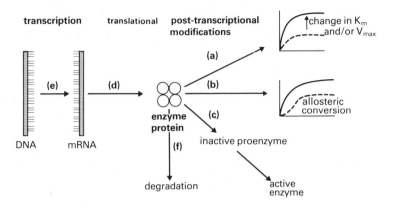

Fig. 4.5 Possible controls of enzyme synthesis during the cell cycle.
a Changing affinity for substrate (manifested by a change of Km and/or Vmax).
b Allosteric conversion of an inactive to active form or *vice versa*.
c Production of active enzymes from inactive proenzyme. Controls **a**, **b**, and **c** change enzyme activity without altering the number of enzyme molecules, and hence operate rapidly.
d Translational control by changing the amount of protein produced.
e Transcriptional control by changing the amount of mRNA produced.
f Change in the rate of degradation of enzyme protein. Controls **d**, **e**, and **f** operate by altering the number of enzyme molecules, and hence work more slowly. (Redrawn from Walker, 1977.)

Table 4.1 Structural differences which affect the control of enzyme synthesis in prokaryotes and eukaryotes

Prokaryotes	Eukaryotes
Genome lies freely within the cell and is available for transcription at any time in the cycle. Transcription and translation are therefore tightly coupled; enzyme synthesis is regulated chiefly at the level of transcription.	Chromosomes are bounded by a nuclear membrane which can act as a barrier for inducers and repressors of gene activity. Therefore, transcription and translation are separated by a regulatory membrane which can allow differential transfer of inducers and repressors across it and can also control the exit of mRNA.
Unstable mRNA molecules are usually produced; the half-life is measured in minutes.	The half-life of the mRNA is measured in hours. Enzymes synthesized continuously may be subject to translational control.
Polyamines may mask areas of the chromosome.	Histones are probably involved in the control of transcription.

The microbial cell cycle

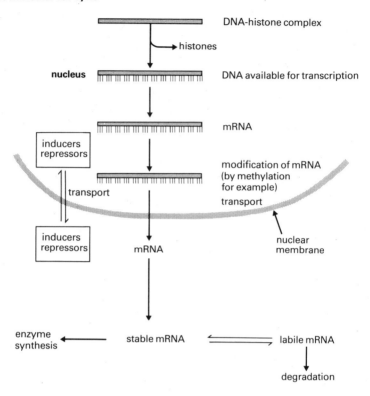

Fig. 4.6 Transcriptional and post-transcriptional controls of enzyme synthesis in eukaryotic cells. DNA is available for transcription only after the removal of associated histones. The mRNA so produced must be modified before it can migrate across the nuclear membrane. Once outside the nucleus the mRNA will eventually be degraded. Inducers and repressors produced in the cytoplasm must be transported across the nuclear membrane to their site of activity. (Redrawn from Walker, 1977.)

The role of cellular proteases Perhaps one of the most neglected aspects of the regulation of enzyme synthesis *in vivo* is the role of intracellular protein degradation. Once synthesized, enzymes are subject to natural degradative reactions—part of the cell's constant turnover of cellular materials. Steady-state levels of enzyme activity reflect not only the rate of synthesis but also the rate of degradation.

A substantial degree of control can be obtained not only by regulating the synthetic process outlined above but also by modulation of the degradative process. Recent studies of protein degradation suggest that a newly synthesized protein has as much chance of being degraded as similar proteins which may have been made some time before: degradation occurs at random (Kay, 1978). Interestingly, it appears that proteins do not have to be defective before they can be degraded. Once initiated, degradation is rapid, and proteins are hydrolyzed to their constituent amino acids rather than to polypeptides of intermediate size.

Individual enzymes probably have characteristic half-lives. These may range from minutes to hours and, in tissues of higher animals, to days. Protein degradation in micro-organisms is not well understood although protease activity in *E. coli* increases in a step-wise fashion during synchronous growth.

Summary

Other possible mechanisms for controlling enzyme activity during the cell cycle include activation and inactivation of enzyme molecules, and transcriptional and translational controls. Eukaryotic micro-organisms are subject to further control imposed by the internal structural complexity of the cell, especially the nucleus. Intracellular proteases may also have an as yet undefined role in modulation of enzyme activity. How all these controls operate during the microbial cell cycle is not yet known.

References

BURDON, R.H. (1976). *RNA biosynthesis*. Chapman and Hall, London.
DONACHIE, W.D. & MASTERS, M. (1969). 'Temporal control of gene expression in bacteria.' In: *The Cell Cycle, Gene-Enzyme Interaction*. Edited by G.M. Padilla, G.L. Whitson and I.L. Cameron. Academic Press, London & New York.
EDWARDS, S.W. & LLOYD, D. (1977). 'Mitochondrial adenosine triphosphatase of the fission yeast *Schizosaccharomyces pombe* 972h⁻.' *Biochemical Journal 162*: 335–345.
FLAVELL, R.A., GLOVER, D.M. & JEFFREYS, A.J. (1978). 'Discontinuous genes.' *Trends in Biochemical Science 3*: 241–244.
FRASER, R.S.S. & NURSE, P. (1978). 'Novel cell cycle control of RNA synthesis in yeast.' *Nature 271*: 726–730.
HALVORSON, H.O., CARTER, B.L.A. & TAURO, P. (1971). 'Synthesis of enzymes during the cell cycle.' *Advances in Microbial Physiology 6*: 47–106.
KAY, J. (1978). 'Intracellular protein degradation.' *Biochemical Society Transactions 6*: 789–797.
WALKER, P.R. (1977). 'The regulation of enzyme synthesis in animal cells.' In: *Essays in Biochemistry* Vol 13. Academic Press, London & New York.

5 Growth, differentiation, and respiration during the microbial cell cycle

Although many cellular processes have been studied during the cell cycles of a variety of micro-organisms, some processes and certain micro-organisms have been investigated in much greater detail than others. We have already examined chromosome replication, largely with reference to *E. coli*, and enzyme synthesis during the cell cycle. In this chapter we shall look at other features of the cell cycles of commonly investigated micro-organisms, some of which have characteristics which make them suitable for the study of particular cellular events.

Prokaryotic differentiation

As we saw in earlier chapters, the cell cycle of prokaryotes is composed of an ordered sequence of events: chromosome replication, septation and cell division. These events usually occur within a cell and the only visible evidence of progression through the cycle is growth in volume and cell division. This visually unexciting progression masks the many temporally linked processes which ensure the fidelity of reproduction of cellular materials in successive generations. However, in some bacteria these processes are closely allied to a series of definite morphological events which occur at specific times in successive cell cycles.

The genus *Caulobacter* affords such an example of simple differentiation which can be used to pin-point known stages in its cycle. The cell cycle of *Caulobacter* has been reviewed by Shapiro (1976), and is presented diagrammatically in Fig. 5.1. A startling feature is the occurrence of two distinct cell types, each having a different morphology. One of the daughter cells arising at division is motile and is propelled by a single polar flagellum. The flagellum is lost after approximately one fifth of the subsequent cycle. This is followed by *de novo* synthesis of a stalk structure at the site of flagellar attachment. As the stalked cell elongates (the stalk being synthesized by extension of the cell wall and membrane) it develops a distinct polarity and a flagellum is synthesized at the pole opposite the stalk. The cell is now known as a *predivisional cell* and it divides asymmetrically by binary fission to give rise to a *swarmer cell* and a *stalked cell*. The two cell types have different morphology but, perhaps more interestingly, they also have different DNA-replication cycles. Stalked cells will begin DNA replication immediately but the

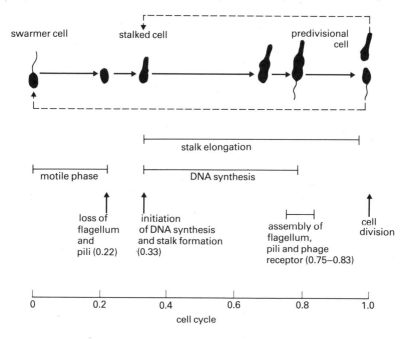

Fig. 5.1 Cell cycle events in *C. crescentus*. The progression and occurrence of the main cycle events are shown during the dimorphic cell cycle, which is represented as a linear scale from 0 to 1.0. The arrows indicate timings of specific processes which occur at known stages (the fractions of the cell cycle during which they occur are shown in parentheses). Dotted lines illustrate the successive cell cycles undergone by each cell type released at division. (Redrawn from Shapiro, 1976.)

swarmer cell must first lose its flagellum and become a stalked cell. It will also be obvious from Fig. 5.1 that the two cell types have cell cycles of different durations.

Another interesting feature of this dimorphic cycle is the time-ordered synthesis of the stalk. This *localized cell wall synthesis* is thought to be confined to the region surrounding the base of the stalk. Little is known about the function of the stalk but, since the bacterium lives in water containing low concentrations of nutrients, it may be a flotation device.

Although the process of cell division is not responsible for the production of dissimilar cell forms (the predivisional cell already shows distinct asymmetry), it may give rise to an event which delays chromosome replication in the swarmer cells. There is evidence that a limited period in the cell cycle for DNA replication must be completed before cell division can occur. The absence of detectable DNA synthesis during the swarmer-cell stage of the cycle is unique to *Caulobacter*.

Our understanding of the temporal events of the *Caulobacter* cell cycle has come from investigations of synchronously dividing cells. Many of the methods used to prepare such cells rely on the different cell morphologies occurring during the cycle. Stalked cells have a greater frictional drag, so by performing a series of differential centrifugations, relatively homogeneous populations can be obtained. Another method of separating cell types exploits the adhesive properties of stalked cells which cling to a variety of surfaces such as glass beads and Petri dishes. One important finding is that cell cycle events always occur in a fixed sequence and at the same stage in the cell cycle, no matter how quickly or slowly the cells are growing.

Much interest has centred on the control and regulation of surface differentiation during the cell cycle of *Caulobacter*. The major landmarks are the synthesis of stalk, flagellum, and pili structures. The stages during the cycle at which they are formed are shown in Fig. 5.1.

Stalk formation can be viewed as a site-specific synthesis of cell wall and membrane material. How is the synthesis of this structure initiated and what controls its elongation? Elongation is affected by the availability of phosphate: high levels of phosphate promote elongation and low levels depress it. Elongation is also controlled by the availability of cyclic GMP, and by the number of generations during which a cell has possessed a stalk. Initiation of stalk formation appears to occur during the series of reactions leading to cell division. In other words, initiation depends on an event which occurs during the previous cell cycle.

The flagellum, pili and phage receptors appear simultaneously at the pole opposite the stalk at approximately 0.7 of the cell cycle. They are also lost simultaneously when the swarmer cell formed at division differentiates into a stalked cell. These structures are synthesized at a definite site on the cell surface and at a definite time. The swarmer cell discards its flagellum on undergoing the swarmer-to-stalk transition. Thus, *Caulobacter* provides an extremely useful system for studying flagellar composition, synthesis, and assembly. In synchronous cultures *flagellin* (the flagellar protein) is synthesized at 0.67–0.73 of a cycle. Its synthesis and subsequent assembly into a visible flagellum is not well understood, but the process must be subject to temporal as well as spatial controls. The production of a flagellum is not essential for cell division since mutants which lack the ability to synthesize flagellin are still able to form polar swarmer cells (possessing only pili and phage receptor sites) which then form stalked cells of the correct polarity after division. In fact, loss of phage receptors or pili does not affect subsequent cell division, DNA replication, or polar stalk formation. It has been demonstrated that RNA synthesis is required throughout the *Caulobacter* cell cycle. The addition of rifampin, an inhibitor of the initiation of RNA synthesis, at different times to synchronous cultures has revealed that RNA synthesis is necessary for loss of swarmer cell mobility, initiation of chromosome replication, and cell division. The use of mutants blocked in the various temporal morphological events of the *Caulobacter* cell cycle has led to the idea that there are two pathways. One pathway follows a series of interdependent events, each of which must be completed before the next one can begin. This pathway comprises DNA replication, cell division, and stalk formation. The other pathway comprises flagellin synthesis and is concomitant with formation of the flagellum, pili, and phage receptor sites. Any step in this second pathway can be omitted without affecting the progression of the other cell cycle events.

Summary

The *Caulobacter* group of bacteria provide a relatively simple system for investigating prokaryotic differentiation. Synchronous cultures are readily prepared and the organism's cell cycle is such that it is highly suitable for investigating the processes of surface growth, initiation of DNA replication, flagellar synthesis and assembly, cell division, and the temporal and spatial expression of surface structures. The genus *Caulobacter* has been included in a group known as *prosthecate bacteria*, members of which also exhibit differentiation during the cell cycle. They have been reviewed by Whittenbury and Dow (1977).

Respiration in prokaryotes

Bacterial respiratory chains have been studied extensively. They vary greatly in composition not only between different species but even in particular organisms. This diversity reflects in part the ability of prokaryotes to adapt very quickly to new environments. Despite the intense interest in bacterial bioenergetics it is surprising how little is known about the development of respiration and oxidative phosphorylation during the bacterial cell cycle. Such studies are also relevant to membrane biogenesis because many of the enzymes associated with bioenergetics—cytochrome oxidase, for example—are intimately associated with membranes. Indeed, they rely on this association for their activity. Progress in the development of respiratory events during the bacterial cell cycle has been reviewed by Poole (1979).

Investigations using synchronous cultures of *E. coli* and *A. eutrophus* prepared by selection using continuous-flow centrifugation have demonstrated that respiration rates increase in a periodic fashion during the cell cycle. In *E. coli*, the pattern of oscillation appears to depend upon the nature of the carbon and energy sources. Oxygen-uptake rates of cells growing with glucose as the carbon source increase to two maxima per cell cycle, whereas the respiratory activity of cells with glycerol as the carbon source increases in two steps per cycle. The mid-points of the increases in respiration rates of the glycerol culture occur at the same stage in the cycle as do the maxima observed in glucose-grown synchronous cultures. Respiration rates in synchronous cultures of *A. eutrophus* growing in the presence of lactate also increase in two steps per cycle.

The pattern of increase in respiration rates during the prokaryotic cycle appears to be expressed as a series of maxima and minima in cells growing with fermentable carbon sources (glucose), and as a stable step-wise increase in cells grown with non-fermentable carbon sources (glycerol or lactate). It is unlikely that the observed periodicities are reflections of changes in respiratory control modulations. They probably reflect the changing demand for energy during the cycle. Processes exerting such a demand might include protein synthesis, chromosome replication and cell division. In the case of *A. eutrophus*, the activity of the enzyme adenosine triphosphatase (ATPase) also changes periodically and in each cycle there are two maxima coincident with the rise in respiratory activity. The ATPase exhibits a change in sensitivity to inhibitors during the cycle indicating profound structural modifications to this important enzyme (Edwards *et al.*, 1978).

Other enzymes involved in the energy-transduction process have rarely been studied. The activity of succinate dehydrogenase, an enzyme usually intimately associated with the cytoplasmic membrane in prokaryotes, increased in a series of steps during the cycle of those bacteria which have been studied. Total NADH dehydrogenase activity increased continuously in a synchronous culture of *B. subtilis* prepared by a selection method. This enzyme was loosely associated with the membrane. In contrast, the membrane-bound enzyme in *Rhodopseudomonas sphaeroides* increased in a step-wise fashion in a synchronous culture prepared by an induction method.

Although we know that observations of enzyme activity during the cell cycle need to be treated with caution, this is not the case when the concentrations of the cytochromes of the respiratory chain are determined. Difference spectroscopy of whole cells or membranes yields direct quantitative determinations of cytochrome content. However, the lack of data is probably due to the relatively large quantity of cell material necessary to make accurate determinations.

This drawback can be overcome by preparing large-scale synchronous cultures or culture fractionation and determining cytochrome concentrations from difference spectra of suspensions at the temperature of liquid nitrogen (77°K). This procedure has the dual advantage that peak heights are enhanced (so less material is needed) and better resolution is achieved, particularly for b and c type cytochromes, which absorb maximally at adjacent wavelengths.

Summary

In the few bacteria so far examined, periodicities in respiratory events such as oxygen uptake, inhibitor sensitivities, and levels of some respiratory enzymes have been observed. Further work in this area may well aid our understanding of bacterial bioenergetics and of the relationship between growth, cell division, and energy production.

Surface extension and the prokaryotic cell cycle

The manner in which cells extend their surfaces in a defined way, form a septum, and maintain cell shape during successive cycles is a baffling problem. How is this achieved in association with other cellular elements during cell growth and division? Normally, an increase in cell volume is accompanied by an extension of the surface area in two rather than three dimensions. For example, rod-shaped bacteria grow in length but not in width. To produce two daughter cells the surface must be redirected in a second dimension at a specific time to form a completed cross-wall at division. Surface extension during the cell cycle of prokaryotes has been reviewed by Daneo-Moore and Shockman (1977) and by Sargent (1978).

Sargent quotes five important facts about growth and formation of rod-shaped bacteria:

> Elongation during balanced growth is continuous but there is no change in width.
>
> At division there may be large variations in the ages of cells.
>
> Variations in size may also occur, but not so widely as variations in age.
>
> Division septa are normally formed at the centre of the cell.
>
> Mean cell length and width depend upon the culture conditions.

How these variations arise and are regulated is not known despite an expanding literature on the topic. Only in streptococci is surface extension and the positioning of the septum well understood.

Shockman and co-workers have proposed that extension of the cell surface in *Streptococcus faecalis* is initiated in a similar fashion to that of chromosome replication in *E. coli*. They propose that, at the same time as the initiation of chromosome replication, a separate series of temporal events is initiated which brings about the synthesis of a complete new unit of cell surface. This series of events, 'a W cycle', takes a fixed time. At generation times of 45 min or longer, a single W cycle occurs per cell cycle. At faster growth rates W cycles are initiated at intervals which depend on the doubling time, so overlapping W cycles can occur. This situation is analagous to the multi-forked chromosome replication observed in rapidly growing cells of *E. coli*. In the same way that multi-forked replication ensures that a complete copy of the *E. coli* genome is inherited by its progeny, so the former process ensures a constant unit length per cell for a given set of growth conditions. It is now widely believed that in bacteria there is a length extension process, punctuated at intervals by septum formation triggered by the chromosome replication cycle. In *E. coli* this event would occur after termination of chromosome replication.

Another aspect of surface extension in bacteria about which little is known is the number and location of cell surface enlargement sites. Various workers using several species of Gram-positive and Gram-negative bacteria have suggested that the process occurs at random at a large number of surface sites, or as a non-random process occurring at a few (say, one, two, or three) defined sites. This may well be an oversimplification. It has been pointed out that the growth of cell surfaces may occur at a fixed number of sites (which are species-dependent) and whose exact number is closely related to the growth rate.

Some models of surface growth in different types of bacteria are illustrated in Fig. 5.2. Only in the case of *S. faecalis* is there good evidence to support the model. These bacteria have between one and three characteristic raised rings of wall material which encircle them at their widest diameter (band J, Fig. 5.2). The band divides, and surface growth occurs between the bands. The resulting bands remain at a constant distance from their respective poles at all stages of the cell cycle. The two J bands serve as the potential division sites for the next generation. The models presented for rod-shaped bacteria and for cocci such as *Staphylococci* which divide in more than one plane are not well supported by experimental evidence. In the case of rod-shaped bacteria the situation is confused by a variety of factors. For example, some Gram-positive rods have a high rate of peptidoglycan turnover which tends to distort patterns of growth.

Investigations into surface extension have been directed towards determining the number and location of growth sites; the rate of surface extension; the timing of appearance of new extension sites;

Growth, differentiation, and respiration during the microbial cell cycle

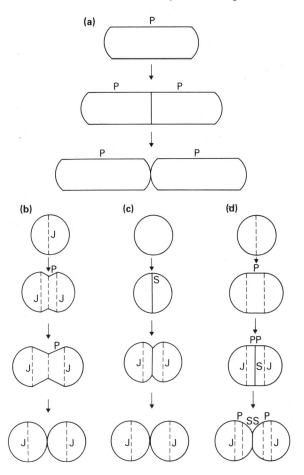

Fig. 5.2 Models of surface growth and septum formation in different types of bacteria.
a In rod-shaped cells peripheral wall (P) is laid down throughout the cell cycle and is punctuated by septum formation (S) at a definite stage of the cell cycle before cell division.
b In *Streptococci* surface extension occurs throughout the cycle, the division sites (J) for the succeeding cycles being located at the boundary formed by new and old cell surfaces.
c In *Staphylococci* the septal wall is laid down early in the cell cycle. A peeling process together with insertion of new wall material throughout the cycle then follows until eventually the two daughter cells fall apart. The bands (J) are again located at the boundary of new and old surface layers and mark the location of the septa formed in the following cell cycle.
d Postulated sequence of events for cocci or short rods in which peripheral wall (P) is initially laid down. This is followed by formation of a septum and a peeling process similar to that described in **c**. (Redrawn from Sargent, 1978.)

The microbial cell cycle

the control of surface growth in relation to the development of other cellular constituents; and the maintenance of cell shape during growth and division.

The approaches used to tackle these problems can be grouped into three main categories: examination of cell size and shape under different environmental conditions; the use of inhibitors; and the use of mutants with abnormal morphologies.

Cell size and shape These are influenced by such factors as the composition of the growth medium, pH, and temperature. A good example is afforded by *Arthrobacter*, which can grow as Gram-positive cocci or Gram-negative rods, depending on the composition of the growth medium.

The use of inhibitors The addition of low concentrations of ß-lactam antibiotics to Gram-negative bacteria effect distinct and characteristic morphological changes. They have revealed the presence of three penicillin-binding proteins in the inner membrane of *E. coli*, each of which has been implicated in a distinct role. Protein 1 is involved in cell elongation, protein 2 in the maintenance of cell shape, and protein 3 in septation. Thus, binding of ß-lactams to protein 1 causes cell lysis; to protein 2 causes changes in cell shape; and to protein 3 inhibition of cell division. Interestingly, inhibition of protein synthesis blocks surface elongation but not peptidoglycan synthesis. The result is a thickening of surface areas but little or no surface growth.

The use of mutants A number of mutants have been characterized which have abnormal morphologies, implying a defect in the process of surface extension and division. Characterization of such mutants may reveal some of the specific controls operating on surface growth, septation, and division.

Summary

Surface growth and septation are closely coupled to macromolecular synthesis during the cell cycle, especially to the chromosome replication cycle. The number and location of sites of surface elongation are well characterized only in streptococci. Environmental factors play a part in maintenance of cell size. Inhibition of protein synthesis stops surface growth, but not peptidoglycan synthesis, which suggests that other factors are involved in surface extension and cell morphology. These may include penicillin-binding proteins, three of which have been implicated in cell elongation, shape maintenance, and septation in *E. coli*.

Eukaryotic cell cycle events: the use of yeasts as experimental material

Yeasts have been used extensively to study a variety of cellular processes primarily because it is easy to cultivate them in large amounts. Some yeasts have also become favoured tools for investigations of cell cycle events. Perhaps the most widely used species is *S. pombe*. Recently, however, *S. cerevisiae* has also been proposed as suitable for investigating the cell cycle (Hartwell, 1974). A number of features make *S. cerevisiae* attractive:

It is single-celled and conveniently cultured in simple media.

Each cell produces a bud which grows until, at division, it forms a daughter cell. Therefore, increase in size of the bud provides an indicator of its progress through the cycle.

A range of well-defined mutants is available.

Cells with chromosome complements ranging from haploid to octaploid may be obtained, all of which *retain the same mitotic cycle*. Thus, DNA synthesis occurs in the same phase of the cycle in polyploids as in haploids. Such cells, therefore, provide a system for studying the effects of gene dosage during the cycle.

The cell cycle of *S. cerevisiae* is shown in Fig. 5.3. It contains a number of distinct landmarks, each of which occurs at a specific time in the cycle. A single unbudded cell has a nucleus containing a single *spindle plaque* embedded in the nuclear membrane, from which microtubules arise. At the end of the G1 period the spindle plaque duplicates and, coincidently, bud emergence and DNA synthesis begin. The bud, once initiated, continues to grow throughout the remainder of the cycle. At the end of the G2 period the nucleus migrates to the neck of the cell and begins the first stages of medial nuclear division. The spindle elongates at this stage. The second stages of nuclear division, called late nuclear division, are followed by *cytokinesis* (cell membrane separation) which leads to actual cell separation and formation of two daughter cells.

Some of these landmarks have been used to assess stages of the cell cycle in synchronously growing cells. Possibly the easiest to identify with the aid of a microscope is bud emergence. Later stages can be assessed from the timing of nuclear division by fixing and staining the cells. The most obvious stage during this period is that when the nucleus is in the neck of the bud. Scoring the number of cells at this point gives a *mitotic index*, which in synchronous cultures will pinpoint a defined stage in the cycle and indicate the degree of synchrony of the culture. In other words, the greater the percentage

The microbial cell cycle

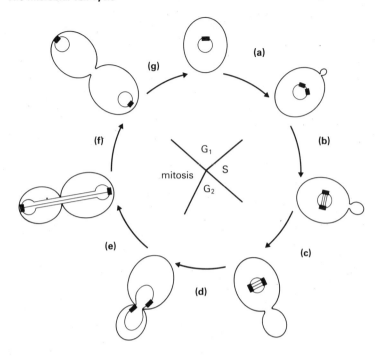

Fig. 5.3 Sequential phases of the *S. cerevisiae* cell cycle.
a The nuclear plaque (shaded box) duplicates and the cell produces a bud at a site on the cell surface.
b and **c** DNA synthesis is initiated and the nuclear plaques separate to opposite poles. The bud continues to grow.
d The nucleus migrates to the neck of the bud to be followed (**e**) by medial nuclear division and spindle elongation such that one nucleus is located in the mother cell and the other in the bud.
f By late nuclear division two distinct nuclear bodies have been formed.
g Once the bud is of approximately the same size as the mother cell, cell separation occurs. In the centre of the figure phases G_1, S, G_2 and mitosis are shown in relation to the other events of the cell cycle. (Redrawn from Hartwell, 1974.)

of cells at this stage at a given time, the higher the degree of synchrony. The easiest way of monitoring synchronous growth of eukaryotes is to count cell numbers. This can be difficult for *S. cerevisiae* because some cells have two or three buds. Generally, cells with one bud are scored as single cells and those with two or three buds as two cells.

The use of yeast *cdc* mutants Recently, studies of the yeast cell cycle have been greatly promoted by the development of *cell-division cycle mutants* (*cdc* mutants). These mutants are usually characterized by dividing normally at one temperature (usually the growth temperature) but show marked changes when transferred to a restrictive temperature (usually a higher temperature). The mutation lies in a gene whose product controls a distinctive discontinuous process in the cell cycle. Mutants are recognized and selected by shifting cells from growth temperature to a restricting temperature and looking for cells blocked in a specific cell-cycle landmark such as bud formation. A large number of these mutants of *S. cerevisiae* have been isolated and characterized. On shifting an asynchronous culture to the restrictive temperature and incubating it for a sufficient time, mutant cells usually assume a characteristic, abnormal morphology. For example, in cells of *S. cerevisiae* carrying the so-called *cdc 13* mutation, the characteristic abnormality occurs in medial nuclear division. After the shift to the restrictive temperature cells complete bud-emergence, DNA synthesis, and nuclear migration, but do not go on to complete medial division, late nuclear division, or cell division. Bud growth continues but no new buds appear on the cell. These *cdc 13* mutants, with large buds and a nucleus located in the cell neck, look different from wild type cells and other non-*cdc* temperature-sensitive mutants.

From the *cdc* mutants isolated so far it has been possible to identify two main groups. The first group includes those mutants which undergo several cell cycles after the shift to the restrictive temperature. The other group consists of cells which have a critical arrest point in the cell cycle. On shifting cultures to the restrictive temperature, cells at stages before the arrest point will become blocked and will not divide, whereas those at stages after this point will undergo division and development will be arrested in the second cycle. The mutants which fall into this second group are said to exhibit *first cycle arrest* on application of the temperature shift.

The isolation and characterization of *cdc* mutants has revealed a temporally ordered sequence of events during the cycle which have been tentatively assigned to a dependent and an independent pathway. The landmarks of the cell cycle (Fig. 5.3) occur in a fixed order which must be integrated and regulated to ensure viable progeny. Use of *cdc* mutants indicates that *certain cell cycle events depend for their initiation on the completion of previous ones*. This coupling ensures a fixed sequence for certain cell cycle events. The sequence of landmarks in the dependent pathway has been determined using inhibitors and *cdc* mutants. A block at a known stage of the cycle causing inhibition or altered expression in subsequent stages implicates the stages as dependent events. If no changes occur then the blocked stage must lie on an independent pathway. Bear in mind that this only holds true as long as the inhibitor or mutation is affecting only one landmark of the cycle.

The microbial cell cycle

The proposed pathways through the cell cycle of *S. cerevisiae* are shown in Fig. 5.4. The events along each pathway depend on the others which occur on that pathway but are independent of events on different pathways. For example, DNA synthesis and nuclear division can occur in the absence of bud emergence and, conversely, bud emergence and nuclear migration occur in the absence of DNA synthesis. The idea of separate pathways during the cell cycle of *S. cerevisiae* is similar to that described for *Caulobacter*. How general this phenomenon is remains to be seen.

Each cell of the yeast *S. pombe* is cylindrical and grows by length extension. The mean diameter of haploid cells of this organism under standard growth conditions is largely invariant at 3.5 μm, while growth in length is from approximately 8 μm to 14 μm. The cell cycle has a number of landmarks. Length extension is not constant; it occurs primarily over the first three-quarters of the cycle, during

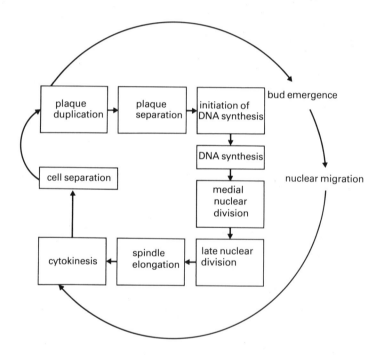

Fig. 5.4 Proposed pathways and major events in the cell cycle of *S. cerevisiae*. The boxed sequence represents the major dependent pathway in which all the events rely on the completion of a previous one. Thus, all the steps of the pathway must be completed in one cycle before a second cycle can commence. The second minor pathway (unboxed) is independent of the first pathway. (Redrawn from Hartwell, 1974.)

which time cell length can be directly related to a cell cycle period. At 0.75 of the cycle extension stops and nuclear division commences, followed at 0.85 of the cycle by the laying down of the central cell plates. The position of the cell plate marks the site of cleavage at division. The size of the cells at nuclear division appears to be under strict control(s). Cells will enter into this process only after they have attained a certain critical size. This will also control the subsequent division.

The fission yeast has been used to study how a cell doubles its complement of cellular constituents during the cycle. This work has been facilitated by the exploitation of mutants which divide at different mean sizes. By comparing patterns of growth and synthesis in a mutant with those known in the wild type, it is hoped to unravel some of the mysteries underlying the way a cell doubles itself during the cycle. Mutants used for this work include a number of *cdc* mutants which divide normally at 25°C but, when shifted to 35°C, are blocked at specific stages of the cycle and do not divide further. Use of such mutants has shown that cell plate formation and cell division depend upon prior nuclear division. Other *cdc* mutants have been isolated which divide at a reduced size at 35°C compared with their size at division at 25°C. All these strains double their rates of polyadenylated messenger RNA synthesis in one step once every cycle. However, the smaller the cell, the later in the cycle this rate doubling occurs. This suggests that the rate of synthesis of polyadenylated messenger RNA is regulated by a size-dependent mechanism which serves to maintain it in constant proportion to the cell mass irrespective of cell size.

Summary

Yeasts have been used to investigate various aspects of the cell cycle. In *S. cerevisiae* this is facilitated by a series of defined events or landmarks—bud emergence, for example—which occur discontinuously during the cell cycle. The use of *cdc* mutants has revealed that some of these events occur in an ordered, highly interdependent manner such that initiation of one stage in the cycle requires the prior completion of another. Studies with *cdc* mutants of *S. pombe* which divide at a reduced size have led to the suggestion that a strict size control operates which allows nuclear division only after a critical cell size has been achieved.

Biogenesis of mitochondria in eukaryotes

The eukaryotic cell is divided into a number of discrete organelles, each of which contains specialized components (DNA in the nucleus, hydrolytic enzymes in the lysosomes, photosynthetic pigments in the chloroplasts and respiratory enzymes in mitochondria). How are these organelles replicated during the cell cycle so that, at division, each daughter cell receives an equal share? Mitochondria and chloroplasts have their own DNA which, in some eukaryotic micro-organisms, may comprise up to 15 to 20% of the total cell DNA. Thus, these organelles may have some degree of autonomy in their biogenesis. Here we will look at the way mitochondria and their associated proteins may be assembled during the cell cycle. The properties of mitochondria in micro-organisms have been discussed extensively (Lloyd, 1974).

We can ask a number of questions about mitochondrial biogenesis:

Are mitochondria formed *de novo* from cellular precursors or by growth and division of existing mitochondria?

Is there a specific time in the cell cycle at which they are synthesized or during which they divide?

How are the various mitochondrial enzymes synthesized during the cycle?

What is the relationship between mitochondrial DNA and nuclear DNA through the cycle. Are they both synthesized at S-phase or is mitochondrial DNA synthesized throughout the cell cycle?

Problems facing the investigator of mitochondrial biogenesis are twofold: first, for a given cell type the number of mitochondria in a cell must be known; second, a convenient method for following the development of mitochondria during the cell cycle is required. Light microscopy and electron microscopy have been used to determine the number of mitochondria per cell. Unfortunately, however, mitochondria are extremely dynamic organelles, constantly moving and changing shape. In addition, a cell is three dimensional, so a count of mitochondria at the cell surface excludes those within.

To overcome this difficulty requires the use of a technique known as *serial sectioning*. A single cell is sliced progressively by micro-manipulation, and photographs are taken of each slice. Mitochondria in each picture are identified and a three dimensional model is reconstructed. The use of this technique has revealed not only the complexity of mitochondrial structure but also how their numbers vary from one cell type to another. Amoebae, for example, may contain up to 3×10^5 mitochondria per cell. Haploid yeasts contain 7 to 17 per cell and diploids 15 to 29 per cell. The unicellular alga *C. reinhardii* contains 9 to 14 mitochondria, but the parasitic blood

Growth, differentiation, and respiration during the microbial cell cycle

flagellates, the trypanosomes, contain only one large mitochondrion which bifurcates throughout the cell to form a complex network. Clearly, due to the time-consuming nature of serial sectioning, this is not a convenient method for following the development of mitochondria during the cell cycle. Current evidence suggests that mitochondria are formed by growth and division of pre-existing organelles. In those cells such as trypanosomes which contain only one mitochondrion which occupies a large volume of the cell, the division cleavage bisects the dividing mitochondrion. Other evidence for the growth and division of pre-existing mitochondria comes from studies of growing fungal hyphal tips which have a zone behind the growing point in which mitochondria divide synchronously.

Little is known regarding the temporal expression of mitochondrial activity during the cell cycle. Some evidence obtained with the alga *C. reinhardii* suggests that small mitochondria join together to form large organelles during certain stages of synchronous growth. This is reflected in changes in oxygen-uptake rates during the cycle. Further evidence has suggested that mitochondria are extremely heterogeneous in make up. Mitochondria from an asynchronous culture of *S. pombe* were separated on a density gradient and the activities of various mitochondrial enzymes assayed. It was noted that the specific activities of different enzymes varied in organelles of different sizes. This could indicate a strict temporal insertion of enzymes into mitochondria during the cell cycle.

There is no suitable direct method for assessing growth and division of mitochondria during the cell cycle but several indirect approaches have been adopted.

Respiratory activity The mitochondrial respiratory chain utilizes oxygen as the final electron acceptor. Measurement of the rate of oxygen uptake by whole cells reflects the respiratory activity of the mitochondria. A number of investigators have determined respiration rates of cells at different stages of synchronous growth. Complex patterns of periodic increases in respiration rates have been observed in many protozoon cell cycles, including *C. fasciculata* and the amoeba *A. castellanii*. Respiratory activity in the yeast *S. pombe* is also periodic during the cell cycle, rising to two maxima in cells growing in the presence of glucose and as two steps in cells growing in the presence of glycerol. This influence of the carbon source on the pattern of respiratory activity is similar to that noted for *E. coli* (page 68).

Enzymes and cytochromes The activities of several mitochondrial enzymes and cytochromes have been assayed during the cell cycles of a number of micro-organisms, in particular *S. pombe*. Some of the enzymes are partially coded for by mitochondrial DNA. For example, four of the nine polypeptides of oligomycin-sensitive ATPase are encoded and synthesized in the mitochondria. Many of the yeast's respiratory enzymes, including cytochrome oxidase and succinate dehydrogenase, are expressed as maxima and minima of activity in cells growing in the presence of glucose, as are cytochromes b and $a + a_3$. In cells growing in the presence of glycerol, the levels of the cytochromes and cytochrome oxidase activity increase in stepwise fashion. The stable and unstable patterns, also exhibited by respiratory activity, possible reflect catabolite repression of respiratory enzymes by glucose (Poole *et al.*, 1974). However, the sequence of events is independent of the carbon source, indicating a strict and ordered assembly of mitochondrial enzymes in *S. pombe*. The pattern of expression of respiratory enzymes in *S. pombe* is dependent to some extent on the composition of the growth medium but their temporal expression is entrained in the cycle.

The relationship between the phases undergone by mitochondrial DNA in relation to those of nuclear DNA is not well understood. Only in the trypanosomes has this been investigated (see Fig. 5.5). The trypanosomes have a large amount of DNA located in the single mitochondrion at a specific region called the *kinetoplast*. Kinetoplast DNA synthesis is tightly coordinated with nuclear DNA synthesis in

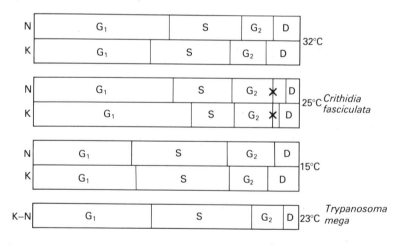

Fig. 5.5 Relationship between the phases G_1, S, G_2, and D undergone by nuclear (N) and kinetoplast (K) DNA in *Crithidia fasciculata* grown at 15°, 25°, and 32°C, and in *Trypanosoma mega* grown at 23°C. **X** marks the onset of mitosis and division of the kinetoplast. (Redrawn from Simpson, 1972.)

C. fasciculata. Depending on the growth temperature, it begins before or just after initiation of nuclear DNA synthesis. This close coupling may be due to a common pool of DNA precursors which occurs only at specific periods of the cell cycle.

Summary

The manner in which eukaryotic micro-organisms sequester an equal share of subcellular organelles at division is not understood. Much work has been directed towards understanding the process of mitochondrial biogenesis; most of the evidence favours independent growth and division of mitochondria. In addition, the periodic expression of many respiratory events indicates a strict temporal control over mitochondrial biogenesis.

References

DANEO-MOORE, L & SHOCKMAN, G.D. (1977). 'The bacterial cell surface in growth and division.' *Cell Surface Reviews 4*: 597–715.
EDWARDS, C., SPODE, J.A. & JONES, C.W. (1978). 'The properties of adenosine triphosphatase from exponential and synchronous cultures of *Alcaligenes eutrophus* H16.' *Biochemical Journal* 172: 253–260.
FANTES, P.A. & NURSE, P. (1978). 'Control of the timing of cell division in fission yeast.' *Experimental Cell Research 115*: 317–329.
HARTWELL, L.H. (1974). '*Saccharomyces cerevisiae* cell cycle.' *Bacteriological Reviews 38*: 164–194.
LLOYD, D. (1974). In: *The mitochondria of micro-organisms.* Academic Press, London & New York.
POOLE, R.K. (1979). 'Temporal diversity of bacterial respiratory systems. Membrane and respiratory development during the cell cycle.' In: *The diversity of bacterial respiratory systems.* Edited by C.J. Knowles. CRC Press Inc., Boca Raton, Florida, USA.
POOLE, R.K., LLOYD, D. & CHANCE, B. (1974). 'The development of cytochromes during the cell cycle of a glucose-repressed yeast, *Schizosaccharomyces pombe* 972h.' *Biochemical Journal 138*: 201–210.
SARGENT, M.G. (1978). 'Surface extension and the cell cycle in prokaryotes.' *Advances in Microbial Physiology 18*: 105–176.
SHAPIRO, L. (1976). 'Differentiation in the *Caulobacter* cell cycle.' *Annual Reviews of Microbiology 30*: 377–407.
SIMPSON, L. (1972). 'The kinetoplast of the Hemoflagellates.' *International Review of Cytology 32*: 139–207.
WHITTENBURY, R. & DOW, C.S. (1977). 'Morphogenesis and differentiation in *Rhodomicrobium vanniellii* and other budding and prosthecate bacteria.' *Bacteriological Reviews 41*: 754–808.

6 Overall summary

The topics covered in the preceding chapters illustrate the diversity of approaches used to investigate the microbial cell cycle. Despite the major differences between prokaryotes and eukaryotes it is possible to draw important conclusions concerning the cell-division cycle in both cell types.

Perhaps the most important point to make is that in asynchronous populations of cells such as batch cultures, the events which together make up the cell cycle are *time-averaged*. Only by investigating cells of similar age can we unravel the progression of cell cycle processes. The controls which superintend them are *temporal*—time-ordered syntheses of cellular components *at specific points* in the cell cycle, and *spatial*—components are synthesized and inserted at *specific cellular locations*. A good example of these controls is that of stalk formation in the prosthecate bacterium *C. crescentus*. This structure is synthesized at a definite *period* of the cell cycle and at a definite *site* on the cell surface.

Recent investigations of the *Caulobacter* and *S. cerevisiae* cell cycles have led to the concept of *sequential cycle units*. Each unit consists of a number of definite events which must be completed before those of the following unit can begin. How universal this model is remains to be elucidated. It will be difficult to investigate in those micro-organisms lacking distinct morphological features expressed at fixed times in the cell cycle. The use of *cdc* mutants should greatly facilitate research in this area.

Although DNA synthesis differs in many major respects between eukaryotes and prokaryotes, some general similarities exist. These include a critical initiation event related to cell size and a possible involvement of repressor or effector proteins. A model has been formulated by Cooper (1979) in an attempt to unify the phases undergone by prokaryotic DNA and eukaryotic DNA during the cell cycle. It is proposed that the phases G1, S, and G2 of eukaryotic DNA are analogous to the I, C, and D-periods during prokaryotic DNA replication. However, there still remains the important difference that eukaryotes, unlike prokaryotes, cannot replicate DNA during cell division.

Another important conclusion that can be drawn is that, in concert with the *DNA-division cycle*, which consists of DNA replication and cell division, there is a *growth cycle* consisting of macromolecular synthesis. There is good evidence for the existence of these two cycles in both eukaryotic and prokaryotic micro-organisms. This stems from

Overall summary

the observation that these cycles can be dissociated from one another by some inhibitors of DNA synthesis.

At present, major areas of research into the cell cycle include investigations of the cell division process in relation to growth and cell size using *cdc* mutants; cell surface growth in bacteria with special reference to maintenance of cell shape and the effects of antibiotics on the surface layers; differentiation processes in relation to the cell cycle using micro-organisms such as prosthecate and budding bacteria, and the slime moulds *D. discoidium* and *P. polycephalum*; respiratory enzymes and ATP production in relation to cell growth and division; and the development of photosynthetic components.

Reliable methods are now available for investigating the cell cycle of a variety of cell types. Possible future areas of research are the testing of pharmaceutical drugs on cells at a known 'sensitive' stage of the cell cycle; characterization of the interaction of plasmid DNA with chromosomal DNA during the cell cycle; and a study of organellar biogenesis in eukaryotes (how, for example, are lysosomes, ribosomes, mitochondria, chloroplasts and peroxisomes synthesized and distributed equally to daughter cells at division?).

Further reading

Review articles

COOPER, S. (1979). A unifying model for the G_1 period in prokaryotes and eukaryotes. *Nature*, *280*: 17–19.

DONACHIE, W.D., JONES, N.C. & TEATHER, R.T. (1973). The bacterial cell cycle. In *Symposium of the Society of General Microbiology* Vol 23 pp.9–44.

HELMSTETTER, C.E., PIERUCCI, O., WEINBERGER, M., HOLMES, M. & TANG, M.S. (1979). Control of cell division in *E. coli*. In *The Bacteria* Vol VII ch 9. Edited by J.R. Sokatch & L.N. Ornston. Academic Press, London & New York.

KOCH, A.L. (1977). Does the initiation of chromosome replication regulate cell division? In *Advances in Microbial Physiology* 16: 49.

LLOYD, D. & TURNER, G. (1980). Structure, function, biogenesis and genetics of mitochondria. In *The eukaryotic microbial cell. Symposium of the Society for General Microbiology*. In press.

PASTERNAK, C.A. (1974). Biochemical aspects of the cell cycle. In *Companion to Biochemistry* ch 12. Edited by A.T. Bull, J.R. Lagnado, J.O. Thomas and K.R. Tipton. Longman, London.

RICHMOND, M.H. (1978). The genetic organization of bacteria and its expression. In *Essays in Microbiology* ch 12. Edited by J.R. Norris and M.H. Richmond. Wiley, Chichester.

Books

BALLS, M. & BILLETT, F.S. (1973). Eds. *The cell cycle in development and differentiation*. Cambridge University Press.

CAMERON, I.L. & PADILLA, G.M. (1966). Eds. *Cell synchrony*. Academic Press, London & New York.

GOLDBERGER, R.F. (1979). Ed. *Biological regulation and development*. Plenum, New York.

MITCHISON, J.M. (1971). *Biology of the Cell Cycle*. Cambridge University Press.

Glossary

Asynchronous culture: a population of randomly dividing cells, such as a batch culture.

Batch culture: cultivation of cells in a growth medium containing a fixed concentration of nutrients. Growth will be exponential for only a few generations and will gradually cease when one or more of the nutrients becomes limiting.

C-period: the period of chromosome replication during the prokaryotic cell cycle.

Cell cycle: the time taken for a cell formed at division to grow and divide.

Cell-division cycle (*cdc*) mutants: mutants which grow and divide normally under standard growth conditions but which become defective in a single cell cycle event when grown at a temperature unfavourable for growth.

Cell initiation mass: prokaryotes exhibit a constancy of size at the initiation of the C-period which has been denoted as Mi—the cell initiation mass.

Continuous culture: continuous cultivation of micro-organisms may be achieved using either *chemostats*, in which fresh growth medium is supplied constantly, culture density being controlled by limiting the concentration of one nutrient, or *turbidostats*, in which all nutrients are supplied continuously in excess but their rate of supply is controlled by culture density.

Continuous enzyme synthesis: synthesis of an enzyme occurs throughout the cell cycle such that enzyme activity doubles in each cycle.

Culture fractionation: separation of asynchronously growing cells into fractions of similarly aged cells. This is usually achieved using density gradient centrifugation.

D-period: period of the prokaryotic cell cycle lasting from the end of chromosome replication to cell division.

DNA-division cycle: the coupled events of DNA replication and cell division but excluding the processes of increase in cell mass (growth).

G_1: period of the eukaryotic cell cycle during which the cell prepares for DNA synthesis.

G_2: period of the eukaryotic cell cycle which lasts from the end of DNA replication to mitosis.

Gene dosage: some genes will be replicated earlier in the cell cycle than others, resulting in a potential doubling of transcriptional and hence translational expression. For example, prokaryotes growing rapidly accommodate chromosome replication cycles by multiple replication forks. Therefore, genes nearest the origin will be present in far greater copy number than those near the terminus.

Glossary

Generation time (g): time taken for any measurable cell component to double during unrestricted exponential growth. It is sometimes also referred to as the *doubling time*.

Growth cycle: those events during the cell cycle which increase cell mass irrespective of DNA replication and cell division.

Histones: basic proteins which are intimately associated with eukaryotic chromosomes. They may prevent transcription of DNA by binding to regions of the chromosomes, and may also have a structural role.

Hypoxia: deprivation of oxygen; hypoxia has been used to induce synchronous growth of micro-organisms.

I-period: the period before initiation of chromosome replication in prokaryotic cell cycles. It may be absent in rapidly dividing cells.

Induction methods: methods used to treat a population of asynchronously dividing cells which induce them to divide synchronously.

Kinetoplast: the single mitochondrion of members of the Trypanosomatidae which contains a high proportion (15–20%) of cellular DNA localized at a specific region of the organelle.

L-forms: mutants of certain Gram-positive and Gram-negative bacteria which cannot synthesize peptidoglycan. They can be cultivated only in media of high osmotic strength.

Linear reading model: the transcription of genes in the order that they occur on the linear eukaryotic chromosomes. This means that the time of synthesis or inducibility of the enzyme is restricted to a definite stage of the cell cycle.

Multiple replication forks: rapidly growing bacterial cells must alter the frequency of initiation of chromosome replication in order to ensure two copies of the genome at division. As a result, bacterial cells can accommodate more than one chromosome replication cycle and multiple replication forks.

Nucleosomes: eukaryotic chromosome sub-units of 150–200 base pairs associated with chromosomal proteins and histones.

Origin: the region of the chromosome at which DNA replication is initiated.

Oscillatory repression model: control of the synthesis of an enzyme by the products of its reaction. This model has been developed mainly for prokaryotes.

Periodic enzyme synthesis: synthesis of an enzyme is restricted to a definite stage in the cell cycle resulting in a periodic rise in activity.

Periodic 'shocks': the application of a perturbing regime at intervals (usually one generation time) to an asynchronous population of cells which eventually brings them to the same phase of the cell cycle.

Polysome: the association of a number of ribosomes with a single mRNA molecule.

Replication fork: the site of replication of double-stranded DNA.

Replicon model: proposes that insertion of new membrane between the

The microbial cell cycle

membrane-chromosome attachment sites serves to segregate the replicated genomes to opposite poles of the cell.

S-phase: period of DNA synthesis during the eukaryotic cell cycle.

Selection methods: preparation of synchronous cultures by the separation of a class of similarly aged cells from an asynchronous culture.

Single 'shock' methods: methods used to induce synchronous growth by the application of a single perturbing treatment to an asynchronous culture, causing the gradual accumulation of cells at a particular stage of the cell cycle.

Synchronous culture: culture consisting of similarly aged cells which grow and divide together.

Synchrony index: measure of the degree of synchrony in a culture; that is, how closely an experimental culture approaches the theoretical one-step growth curve.

Temperature sensitive (*ts*) mutants: mutants which grow normally at the optimum growth temperature but which lose the activity of a particular protein at a higher or lower temperature.

Termination proteins: proteins which are believed to be synthesized at or after the termination of bacterial chromosome replication and which have a function in the ensuing division process.

W-cycles: model for the initiation of a complete unit of cell surface proposed for *S. faecalis*. In rapidly growing cells overlapping W-cycles are initiated rather like the overlapping chromosome replication cycles observed in rapidly growing *E. coli*.

Index

Acanthamoeba castellanii 21, 43, 79
Alcaligenes eutrophus 21, 35, 68
amino acid auxotrophs 10
Amoeba proteus 41
Arthrobacter 72
Astasia longa 13
asynchronous 2
Bacillus 9, 23
Bacillus subtilis 3, 9, 26, 32, 35–37, 58, 68
balanced growth 7
batch culture 1, 82
bidirectional chromosome
    replication 27–30
budding during the cell cycle
    of yeast 73–77
C-period 26–30, 35, 38–39, 82
Candida utilis 21
Caulobacter crescentus 3, 25, 64–67, 76, 82
cell cycle 1, maps, 7–8
cdc mutants 74–77, 82–83
cell envelope and chromosome
    segregation in prokaryotes 35–38, 40
cell septation 39–40
centrifugal elutriation 21–22
centrifugation 15
    continuous-flow 19–21, 68
    counter-flow 21–22
    for selection synchrony 15–19
chemostats 1, 12
Chlamydomonas reinhardii 14, 78–79
Chlorella 4, 58
Chlorella pyrenoidosa 41, 55
Chlorella vulgaris 4, 14
chromosome replication
    in prokaryotes 26–30
    initiation of 30–34
    role of ATP in 34–35
    segregation of 35–38, 40
    cell division and 38–40
    replication, initiation in
        eukaryotes 40–45
continuous-flow centrifugation 19–21, 23
Crithidia 45
Crithidia fasciculata 19, 51, 79–81
Crithidia luciliae 11
culture fractionation 2, 5, 22–23, 69
    using density gradients 22–23
    by continuous-flow centrifugation 23
Cylindrotheca fusiformis 10
cytochromes 79–80
cytokinesis 73–74

D-period 27–30, 35, 38–39, 82
density gradients 17–19, 22–23, 37
deoxyadenosine 11–12
dextran 19
Dictyostelium discoidium 25, 82
differentiation 25
    in prokaryotes 64–67
discontinuous genes 60
DNA division-cycle 11–12
    in prokaryotes 25–40, 48, 82–83
    in eukaryotes 40–47
DNA, extrachromosomal 41, 44–45
DNA replication unit 41–43
DNA synthesis, inhibitors of 11–12, 82–83
end points of growth 9
enzyme potential 3
enzyme synthesis and the cell cycle 3, 51–55
    controls of 60–63
    models of 55–60
    patterns of 54–55
    repression and de-repression of 53–60
Escherichia coli 1, 4, 9, 12, 15–17, 19–21, 25–27, 29–32, 35–40, 43, 47, 51, 53, 55–59, 63, 64, 68, 70, 72, 79
Euglena gracilis 13–14
ficoll 9
filters, membrane 16–17, 27–30
filtration
    methods for cell size selection 15–17
first cycle arrest 75
flagellin 66
flagellum 65–67
G_0 40
G_1 40–41, 73–74, 80, 82
G_2 10, 40–41, 73–74, 80
ß-galactosidase 3, 53–58
gene dosage 30, 58
generation time 6
ß-glucosidase 59–60
growth cycle 11, 38, 48, 82
histones 40, 44, 48, 61–62
hydroxyurea 11, 44
hypoxia 10–11, 14
I-period 26–30, 35, 82
inducers of enzyme synthesis 54–55, 59–60, 62
induction methods for synchrony 7–14
inhibitors
    of DNA synthesis 11–12, 14
    of surface growth 72

Index

initiation of chromosome replication see chromosome
initiation mass 30, 35
kinetoplast 45, 80
L-forms 36
lac operon 56–58
ß-lactam antibiotics 39, 72
landmarks of the cell cycle in yeast 73–77
Leishmania 45
light-dark cycles 13–14
linear reading model 59–60
membrane and chromosome replication 31–32, 36–38
Mi 30, 32
micronuclei 41, 44
mitochondria, biogenesis of 77–81
  DNA in 45, 50, 78
mitotic cycle 45, 73
  index 73
nalidixic acid 35, 39
negative feedback
  of enzyme synthesis 58
nucleolus 40, 44, 50
nucleosomes 41
operator 56–59
origin, of replication 27, 30, 35–36
oscillatory repression model 57–59
palindromic sequences 44
Paramecium aurelia 41, 44
penicillin-binding proteins 72
periodic enzyme synthesis 53–55, 59
periodic shocks 14
periodic supply of growth factors 12
Physarum polycephalum 24, 43–44, 82
pili 65–67
plasmodium 24
polyamines 61
polycistronic 50
polysome 50
Polytomella agilis 10
predivisional cell 64–67
prosthecate bacteria 67
proteases 62–63
protein
  as effector/repressor of C 32–34
  synthesis during the cell cycle 52–55
Proteus vulgaris 9
replication fork 27, 35
  multiple forks 30
replicon 41
  model 31–32, 36
respiration during prokaryotic cell cycle 67–69
  activity during the eukaryotic cell cycle 79–80

Rhodopseudomonas palustris 3
Rhodopseudomonas sphaeroides 68
RNA-polymerase 51, 59
  synthesis during the cell cycle 11, 48–52, 67
mRNA 48–52, 58, 60–62, 77
  polyadenylated 51, 77
  polycistronic 50, 57
rRNA 44–45, 48–52
tRNA 48–52
S-phase 3, 11–12, 40–47, 73–74, 79–80, 82
Saccharomyces cerevisiae 4, 22, 41, 53, 73–77, 82
Saccharomyces dobzhanskii 59
Saccharomyces fragilis 59
Salmonella typhimurium 13, 57
Schizosaccharomyces pombe 2, 4, 11, 19, 21, 23, 41, 51, 53, 55, 60, 73, 76–77, 79–80
selection methods 7, 15–22
sequential cycle units 82
serial sectioning 78
single shock methods 14
spindle plaque 73–74
spores, synchronous germination of 10, 23
stalked cell 64–67
Staphylococci 70–71
Streptococci 69–72
Streptococcus faecalis 70
sucrose 18–19
surface extension in prokaryotes 69–72
swarmer cell 3, 64–67
synchronous cultures 2
  methods for preparation of 5, 8–22
synchrony
  assessment of 5–6
  index 5–6
  naturally occurring 5, 23–24
temperature changes for induction of synchrony 13–14
temperature sensitive (ts) mutants 34
termination proteins 38
terminus 27
Tetrahymena pyriformis 4, 10, 13, 21, 41, 44–45
transcription 48, 59–62
transition point 47
translation 48, 59–62
tritiated thymidine 27–30, 43
trypanosomes 79
turbidostat 1
W-cycle 70
zonal rotors 19, 22–23